AF522291

Active Transport in Plants

B.Poornima

RANDOM PUBLICATIONS
NEW DELHI (INDIA)

Active Transport in Plants

ISBN 978-93-5111-319-5

Published in 2014 in India by

RANDOM PUBLICATIONS

4376-A/4B, Gali Murari Lal, Ansari Road
New Delhi-110 002
Phone : +91-11-43580356, +91-11-23289044
e-mail: randomexports@gmail.com, sales@randompublications.com, info@randompublications.com
Reprinted 2018

Type Setting by : Keystoneprintads, Delhi-110051
Digitally Printed at: Replika Press Pvt. Ltd.

Preface

Active transport is the movement of all types of molecules across a cell membrane against its concentration gradient (from low to high concentration). In all cells, this is usually concerned with accumulating high concentrations of molecules that the cell needs, such as ions, glucose and amino acids. If the process uses chemical energy, such as from adenosine triphosphate (ATP), it is termed primary active transport.

Plants are classified as autotrophs because they manufacture their needed nutrients by photosynthesis, converting carbon dioxide and water to sugar fuels with the addition of energy from the Sun. In times of rapid photosynthesis, the main product is glucose, but it is usually converted to the larger sugar sucrose.

These sugars that are synthesized in the leaves must be transported to other parts of the plant. Other structures in the plants such as roots and flowers require the energy but cannot manufacture it. Also, sugars may be stored in the roots and stem.

Sometimes organisms require certain substances and therefore need to move dissolved substances from a region of low concentration to a region where they are at a higher concentration.

This process is of course the opposite to the direction in which particles would normally move in diffusion. This is active transport. In active transport particles move against a concentration gradient and therefore require energy which must be supplied by the cell.

The movement of a substance across a membrane from a region of its lower concentration to a region of its higher concentration against a concentration gradient, using energy. This energy is supplied through respiration using ATP. Mitochondria (cell organelles in the cytoplasm) control energy release. Respiratory poisons block energy release, so they can prevent active transport.

Present book has been complete to cater to the requirements of the teachers, researchers and students.

I thank all members of my team who have helped in the preparation of the book. My special thanks go to "Random Publications" who have published the book.

—B.Poornima

Contents

1

Transport in Plants

INTRODUCTION

Two main types of plant tissue are used in transport - xylem and phloem. Xylem transports water and minerals. Phloem transports organic molecules such as the products of photosynthesis.

XYLEM

There are four types of xylem cells:

- Xylem vessels: Consist of dead hollow cells because the walls are lignified and the cell contents disintegrate. The lignin makes the cell wall impermeable so they are in effect waterproof. It also makes the vessels extremely strong and prevents them from collapsing. They have a wide lumen and are linked end to end to create a long, hollow tube since the end cell walls have one or many perforations in them. This allows the transport of large volumes of water. The sidewalls have bordered pits (unlignified areas) to allow lateral movement of water. Xylem vessels are found in angiosperms.
- Tracheids: Similar to vessels but with narrower lumens and connected by pits. They have tapered ends so that they dovetail together. Tracheids are found in conifers.
- Parenchyma: Living cells with thin cellulose walls. They can store water, which makes them turgid and so gives them a supporting role.
- Fibres: They provide strength because their walls are lignified (and therefore, dead).

MOVEMENT IN THE ROOT

Water enters through the root hair cells and then moves across into the xylem tissue in the centre of the root. Water moves in this direction because the soil water has higher water potential, than the solution inside the root hair cells. This is because the cell sap has organic and inorganic molecules dissolved in it. The root hairs provide a large surface area over which water can be absorbed. Minerals are also absorbed but, as you should be able to work out, their absorption requires energy in the form of ATP because they are absorbed by active transport. They have to be pumped against the concentration gradient.

Water taken up by the root hairs moves across the cortex of the root either via the cytoplasm of the cells in between the root hair cell and the xylem (the symplast pathway) or through the cell walls of these cells (theapoplast pathway). The root hair cell will have higher water potential than the cell next to it. As always, water moves by osmosis to where the water potential is lower. In this way, as water is always being absorbed by the root hairs, water will always move towards the centre of the root.

When the water reaches a part of the root called the endodermis, it encounters a thick, waxy band of suberinin the cell walls. This is the Casparian strip and it is impenetrable. In order to cross the endodermis, the water that has been moving through the cell walls must now move into the cytoplasm.

Once it has moved across the endodermis, it continues down the water potential gradient until it reaches a pit in the xylem vessel. It enters the vessel and then moves up towards the leaves.

MOVEMENT IN THE XYLEM

Water evaporates from the mesophyll cells into air spaces in the leaf. If the air surrounding the leaf has less water vapour than the air in the intercellular spaces, water vapour will leave the leaf through stomata.

This process is called transpiration and will continue as long as the stomata are open and the air outside is not too humid. On dry, windy days when water vapour is continually diffusing out and being removed, transpiration will increase in rate. Although this loss of water can cool the plant, it is essential that the plant does not lose too much water. Therefore water must be continuously supplied to the leaves. The xylem ensures that this happens. Xerophytes are plants which are well adapted to living where conditions are very dry.

They may have rolled up leaves - for example, Marram grass which exposes the waterproof cuticle on the outside and means the stomata open into an inner humid space. Other Xerophytes store water in their stems and reduce the surface area of their leaves, which become spines - for example, Cactus.

Water is removed from the top of xylem vessels into the mesophyll cells down the water potential gradient. This removal of water from the xylem reduces the hydrostatic pressure exerted by the liquid so the pressure at the top is less than at the bottom. This pushes the water up the tube. The surface tension of the water molecules, the thin lumen of the xylem vessels and the attraction of the water molecules for the xylem vessel wall(adhesion), helps to keep the water flowing all the time and to keep the water column intact.

Pressure to push water up can also be increased from the bottom. By actively pumping minerals from cells surrounding the xylem into the xylem itself, more water is drawn into the xylem by osmosis.

This increase in water pressure, called root pressure, certainly helps in the process but is less important than the simple movement of water down the water potential gradient, ultimately from the soil at the bottom, to the air at the top. This is because moving water this way does not require energy (it is passive).

PHLOEM

There are four types of phloem cells:

- Sieve tube elements: These are living, tubular cells that are connected end to end. The end cell walls have perforations in them to make sieve plates. The cytoplasm is present but in small amounts and in a layer next to the cell wall. It lacks a nucleus and most organelles so there is more space for solutes to move. The cell walls are made of cellulose so solutes can move laterally a well as vertically. Next to each sieve tube element is a companion cell.
- Companion cell: Since the sieve tube element lacks organelles, the companion cell with its nucleus, mitochondria, ribosomes, enzymes etc., controls the movement of solutes and provides ATP for active transport in the sieve tube element. Strands of cytoplasm called plasmodesmata connect the sieve tube element and companion cell.
- Parenchyma: Provides support through turgidity.
- Fibres: Provides support for the sieve tube elements.

MOVEMENT IN THE PHLOEM

This process is called translocation and involves the movement of organic substances around the plant. It requires energy to create a pressure difference and so is considered an active process.

Sucrose is loaded into the phloem at a source, usually a photosynthesizing leaf. For this to occur, hydrogen ions are pumped out of the companion cell using ATP. This creates a high concentration of hydrogen ions outside the companion cell. Sucrose is loaded (moved into companion cells) by active transport, against the concentration gradient.

However, the protein carrier involved in the loading, has two sites, one for sucrose and one for a hydrogen ion. When it is used to pump sucrose into the companion cell, hydrogen will move in the opposite direction, back down its concentration gradient. This is why a high concentration of ions is needed outside the cell.

The sucrose can then diffuse down the concentration gradient into the sieve tube element via the plasmodesmata that connects the companion cell with the sieve tube element. This lowers the water potential of the sieve element so water enters by osmosis.

At another point sucrose will be unloaded from the phloem into a sink (e.g. root). It is likely that the sucrose moves out by diffusion and is then converted into another substance to maintain a concentration gradient. Again, water will follow by osmosis.

This loading and unloading results in the mass flow of substances in the phloem. There is evidence to support this theory; the rate of flow in the phloem is about 10,000 times faster than it would be if it was due only to diffusion, the pH of the phloem sap is around 8 (it is alkaline due to loss of hydrogen ions), and there is an electrical potential difference across the cell surface (negative inside due presumably to the loss of positively charged ions).

ACTIVE TRANSPORT

Active transport is the movement of all types of molecules across a cell membrane against its concentration gradient (from low to high concentration). In all cells, this is usually concerned with accumulating high concentrations of molecules that the cell needs, such as ions, glucose and amino acids. If the process uses chemical energy, such as from adenosine triphosphate (ATP), it is termed primary active transport. Secondary active transport involves the use of an electrochemical gradient. Active transport uses cellular energy, unlike passive transport, which does not use cellular energy. Active transport is a good example of a process for which cells require energy. Examples of active transport include the uptake of glucose in the intestines in humans and the uptake of mineral ions into root hair cells of plants.

DETAILS

Specialized transmembrane proteins recognize the substance and allow it access (or, in the case of secondary transport, expend energy on forcing it) to cross the membrane when it otherwise would not, either because it is one to which the phospholipid bilayer of the membrane is impermeable or because it is moved against the direction of the concentration gradient. The last case, known as primary active transport, and the proteins involved in it as pumps, normally uses the chemical energy of ATP. The other cases, which usually derive their energy through exploitation of an electrochemical gradient, are known as secondary active transport and involve pore-forming proteins that form channels through the cell membrane.

Sometimes the system transports one substance in one direction at the same time as cotransporting another substance in the other direction. This is called antiport. Symport is the name if two substrates are being transported in the same direction across the membrane. Antiport and symport are associated with secondary active transport, meaning that one of the two substances is transported in the direction of its concentration gradient utilizing the energy derived from the transport of second substance (mostly Na+, K+ or H+) down its concentration gradient.

Particles moving from areas of low concentration to areas of high concentration (i.e., in the opposite direction as the concentration gradient) require specific trans-membrane carrier proteins. These proteins have receptors that bind to specific molecules (e.g., glucose) and thus transport them into the cell. Because energy is required for this process, it is known as 'active' transport. Examples of active transport include the transportation of sodium out of the cell and potassium into the cell by the sodium-potassium pump. Active transport often takes place in the internal lining of the small intestine.

Plants need to absorb mineral salts from the soil or other sources, but these salts exist in very dilute solution. Active transport enables these cells to take up salts from this dilute solution against the direction of the concentration gradient.

PRIMARY ACTIVE TRANSPORT

Primary active transport, also called direct active transport, directly uses energy to transport molecules across a membrane.

Most of the enzymes that perform this type of transport are transmembrane ATPases. A primary ATPase universal to all life is the sodium-potassium pump, which helps to maintain the cell potential. Other sources of energy for Primary active transport are redox energy and photon energy (light). An example of primary active transport using Redox energy is the mitochondrial electron transport chain that uses the reduction energy ofNADH to move protons across the inner mitochondrial membrane against their concentration gradient. An example of primary active transport using light energy are the proteins involved in photosynthesis that use the energy of photons to create a proton gradient across the thylakoid membrane and also to create reduction power in the form of NADPH.

MODEL OF ACTIVE TRANSPORT

ATP hydrolysis is used to transport hydrogen ions against the electrochemical gradient (from low to high hydrogen ion concentration). Phosphorylation of the carrier protein and the binding of a hydrogen ion induce a conformational (shape) change that drives the hydrogen ions to transport against the electrochemical gradient. Hydrolysis of the bound phosphate group and release of hydrogen ion then restores the carrier to its original conformation.

ATP USING PRIMARY ACTIVE TRANSPORT TYPES

1. P-type ATPase: sodium potassium pump, calcium pump, proton pump
2. F-ATPase: mitochondrial ATP synthase, chloroplast ATP synthase
3. V-ATPase: vacuolar ATPase
4. ABC (ATP binding cassette) transporter: MDR, CFTR, etc.

SECONDARY ACTIVE TRANSPORT

In secondary active transport, also known as *coupled transport* or *co-transport*, energy is used to transport molecules across a membrane; however, in contrast to primary active transport, there is no direct coupling ofATP; instead, the electrochemical potential difference created by pumping ions out of the cell is used. Permitting one ion or molecule to move from the side where it is more concentrated to that where it is less concentrated increases entropy and can serve as a source of energy for metabolism (e.g. in ATP synthase). In August 1960, in Prague, Robert K. Crane presented for the first time his discovery of the sodium-glucose cotransport as the mechanism for intestinal glucose absorption. Crane's discovery of cotransport was the first ever proposal of flux coupling in biology.

Cotransporters can be classified as symporters and antiporters depending on whether the substances move in the same or opposite directions.

ANTIPORT

In an antiport two species of ion or other solutes are pumped in opposite directions across a membrane. One of these species is allowed to flow from high to low concentration which yields the entropic energy to drive the transport of the other solute from a low concentration region to a high one. An example is the sodium-calcium exchanger or antiporter, which allows three sodium ions into the cell to transport one calcium out.

Many cells also possess a calcium ATPase, which can operate at lower intracellular concentrations of calcium and sets the normal or resting concentration of this important second messenger. But the ATPase exports calcium ions more slowly: only 30 per second versus 2000 per second by the exchanger. The exchanger comes into service when the calcium concentration rises steeply or "spikes" and enables rapid recovery. This shows that a single type of ion can be transported by several enzymes, which need not be active all the time (constitutively), but may exist to meet specific, intermittent needs.

SYMPORT

Symport uses the downhill movement of one solute species from high to low concentration to move another molecule uphill from low concentration to high concentration (against its electrochemical gradient). Both molecules are transported in the same direction.

An example is the glucose symporter SGLT1, which co-transports one glucose (or galactose) molecule into the cell for every two sodium ions it imports into the cell. This symporter is located in the small intestines, trachea, heart, brain, testis, and prostate. It is also located in the S3 segment of the proximal tubule in each nephron in the kidneys. Its mechanism is exploited in glucose rehydration therapy and defects in SGLT1 prevent effective reabsorption of glucose, causing familial renal glucosuria.

EXAMPLES

- Metal ions, such as Na′ K′ Mg′ or Ca′ require ion pumps or ion channels to cross membranes and distribute through the body
- The pump for sodium and potassium is called sodium-potassium pump or Na /K -ATPase
- In the epithelial cells of the stomach, gastric acid is produced by hydrogen potassium ATPase, an electrogenic pump
- Water, ethanol, and chloroform exemplify simple molecules that do NOT require active transport to cross a membrane.

ENDOCYTOSIS

Endocytosis is the process by which cells take in materials. The cellular membrane folds around the desired materials outside the cell. The ingested particle becomes trapped within a pouch, vacuole or inside the cytoplasm. Often enzymes from lysosomes are then used to digest the molecules absorbed by this process. Biologists distinguish two main types of endocyctosis: pinocytosis and phagocytosis.

- In pinocytosis, cells engulf liquid particles (in humans this process occurs in the small intestine, cells there engulf fat droplets).
- In phagocytosis, cells engulf solid particles.

TRANSPORT OF MOLECULES ACROSS CELL MEMBRANES

DIFFUSION AND SELECTIVITY

Transport proteins play a critical part in solute movement because membranes constitute a barrier to free diffusion of molecules. If membranes allowed rapid diffusion of both ions and water, gradients of ions would be slight and osmotic pressure could only be achieved with enormous energy consumption. Instead, some molecules, such as water and gases, move rapidly across membranes. Unrestricted movement of water relative to solutes is the basis of osmosis, and in plants the generation of *P*. Similarly, if the movement of CO_2 and O_2 to and from the sites of photosynthesis and respiration were substantially impeded, steep concentration gradients would reduce the efficiency of these vital processes. The principles of *diffusion* and *selectivity*, which are used to describe differential rates of molecular movement, provide a physical rationale for osmosis.

In a homogeneous medium, net movement of molecules down their concentration gradient is described by Fick's First Law of diffusion. The molecule and medium may be a solute in water, a gas in air or a molecule within the lipid bilayer; a version of this equation describing water movement in soil. Fick's Law holds when the medium is homogeneous in all respects except for the concentration of the molecule. If there was an electric field or a pressure gradient then Fick's Law may not be appropriate. Considering the case of a solute in water, say, sugar, Fick's Law states that net movement of this solute, also called the net flux (J_s), is proportional to the concentration gradient of the solute ÄC_s/Äx:

The diffusion coefficient (D_s) is a constant of proportionality between flux, J_s, and concentration gradient (mM m), where solute concentration (C_s) varies over a distance (Äx). Flux is measured as moles of solute crossing a unit area per unit time (mol m s), so D_s has the units m s. D_s has a unique value for a particular solute in water which would be quite different from D_s for the same solute in another medium, for example the oily interior of a lipid membrane.

When we consider diffusion of a molecule across a membrane from one solution to another, Fick's Law can be applied to each phase (solution 1–membrane–solution 2). However, flux across the membrane also depends on the ability of the molecule to cross boundaries (i.e. to partition) from solution into the hydrophobic membrane and then from the membrane back into solution. Another difficulty is that the thickness of membranes is relatively undefined and we need to know this for Fick's equation above (Äx). The two solutions might differ in pressure and voltage and these can change steeply across a membrane; however, if for simplicity we consider a neutral solute at low concentration, these factors are not relevant. A practical quantitative description of the flux of neutral molecules across membranes uses an expression intuitively related to Fick's Law stating that flux across a membrane (J_s) of a neutral molecule is proportional to the *difference in concentration* (ÄC_s):

The constant of proportionality in this case is the permeability coefficient (P_s), expressed in m s. When P_s is large, solutes will diffuse rapidly across a membrane under a given concentration gradient. P_s embodies several factors: partitioning between solution and membrane, membrane thickness and diffusion coefficient of the solute in the membrane. It can be used to compare different membranes and to compare treatments that might alter the ability of a solute to move across the membrane. Note that Equation 4.2 assumes that the *membrane limits the rate of solute flux* and that concentration gradients leading to diffusion in solutions adjacent to the membrane will not be significant. If the two solutions are stirred rapidly then this will help to justify this assumption. However, there is always an unstirred layer adjacent to the membrane through which diffusion occurs, and for molecules that can permeate the membrane very rapidly the unstirred layer can be a problem for the correct measurement of permeability.

P_s differs markedly for different molecules passing through lipid-based membranes. Permeabilities can differ by eight orders of magnitude, reflecting the selectivity of native lipid bilayers. Differences in membrane permeabilities between molecules are much larger than differences in diffusion coefficients in free solution, the latter depending on the size of molecules rather than membrane properties. Collander (1954) showed that partitioning between water and oil phases (expressed as partition coefficient) was a determining factor for membrane permeability. Charged molecules and large polar molecules do not easily partition into the oily membrane while some polar molecules such as water, methanol and urea permeate faster than predicted from their partition coefficients. This indicates that there are special pathways for the movement of these small molecules other than through the lipid phase. A comparison of artificial lipid membranes with biological membranes supports this notion because it shows that many molecules and ions permeate biological membranes much faster than would be predicted on the basis of oil solubility and size. For these solutes there are *transport proteins* in biological membranes that increase solute permeability.

The reflection coefficient usually ranges between zero and one. Some substances such as mercuric chloride decrease water permeability of plant membranes so much that s becomes negative. Close consideration of *P* in the presence of different solutes reinforces the importance of s for osmosis. Using a pressure probe to measure *P*, the membrane is found to be ideally semipermeable for sucrose (s = 1); that is, the membrane almost totally 'reflects' sucrose. Over long periods, sucrose is taken up slowly but permeability relative to water is negligible. In this case, the change in *P* would be equivalent to the change in P. If s is near zero, then water and the solute (say, propanol) are equivalent in terms of permeability. No change in *P* can be generated across a cell wall if s is zero. If s is negative, this leads to the intriguing situation where *P* changes in the opposite way to that which we normally expect.

For plant and animal cells, most major osmotic solutes have a s of about one so Equation 4.3 can be presented without s as a variable. However, s could be important when highly permeable solutes like ethanol, urea and ammonia reach significant levels, for example when membranes of waterlogged plants are exposed to ethanol and ammonia.

CHEMICAL POTENTIAL

Diffusion of neutral molecules at low concentrations is driven by differences in concentrations across membranes. There are other forces that may influence solute diffusion, including the voltage gradient when considering movement of charged molecules (ions) and the hydrostatic pressure when considering movement of highly concentrated molecules (such as water in solutions). These forces can be added to give the total potential energy of a particular molecule relative to a reference value:

Total potential = Reference potential + Concentration + Electrical + Pressure. Gravitational potential energy could also be added to this equation if we were to examine the total potential over a substantial height difference, but for movement of molecules across membranes this is not relevant. The formal relationship for the chemical potential of a molecule *j* (u_j), measured as energy content per mole (joules mol) and using the same order of terms as the expression.

The *concentration term* is a measure of the effect on chemical potential of the concentration (actually the activity which is usually somewhat less than total concentration). The gas constant, *R* (8.3143 joules mol K), and absolute temperature, *T*(equals 273.15 plus temperature in degrees Celsius, expressed in degrees Kelvin), account for the effects of temperature on chemical potential. Incidentally, from this term and the pressure term, the well-known van 't Hoff relation can be derived for osmotic pressure: P = *RTC*.

The electrical term is a measure of the effect of voltage (*E*) on chemical potential. The charge on a solute (*z*) is obviously relevant since if it was zero this term would not contribute to the total potential. The sign also determines whether an ion is repelled or attracted by a particular voltage. Electrical charge and concentration are related by the Faraday constant (*F*) which is 96 490 coulombs mol. The electrical and concentration terms form the basis of the Nernst equation.

The *pressure term* measures the effect of hydrostatic pressure on chemical potential, where *P* = pressure and $_s$ is the partial molar volume of the solute.

Molecules diffuse across a membrane down a chemical potential gradient, that is, from higher to lower chemical potential. Diffusion continues until the difference in chemical potential equals zero, when equilibrium is reached. The direction of a chemical potential gradient relative to transport of a molecule across that membrane is important because it indicates whether energy is or is not added to make transport proceed. Osmotic 'engines' must actively pump solutes against a chemical potential gradient across membranes to generate *P* in a cell. Sometimes ions move against a concentration gradient even when the flux is entirely passive (no energy input) because the voltage term dominates the concentration term in Equation 4.5. In this case, ions flow according to gradients in electrical and total chemical potential. For this reason, the chemical potential of ions is best referred to as the *electrochemical potential*.

IONS, CHARGE AND MEMBRANE VOLTAGES

Ions such as potassium and chloride (K and Cl) are often major osmotic solutes in plant cells. In fact, deficiency of either of these two inorganic nutrients can increase a plant's susceptibility to wilting. In addition, most other inorganic nutrients are acquired as ions and some major organic metabolites involved in photosynthesis and nitrogen fixation bear a charge at physiological pH. For example, malic acid is a four-carbon organic acid that dissociates to the divalent malate anion at neutral pH. Besides the role in generation of *P*, ionic fluxes and the associated electrical effects of these fluxes are components of signalling in plants. Calcium (Ca) fluxes across cell membranes are involved in cell signalling and although not osmotically significant they play a crucial role in the way cells communicate and self-regulate. Finally, some ions are used to store energy but need not occur at osmotically significant concentrations. Cell membranes from all kingdoms use hydrogen (H) ions (protons) in one way or another to store energy that can be used to move other ions or to manufacture ATP. The highest concentration of H that occurs is only a few millimoles per litre yet H plays a central role in energy metabolism. Other ions such as sodium (Na) can also be used to store energy in plant cells.

To understand ion movement across membranes, two crucial points must be understood: (1) ionic fluxes alter and at the same time are determined by voltage across the membrane; (2) in all solutions bounded by cell membranes, the number of negative charges is balanced by the number of positive charges. Membrane potential is attributable to a minute amount of charge imbalance that occurs on membrane surfaces. So at constant membrane potential the flux of positive ions across a membrane must balance the flux of negative ions. Most biological membranes have a capacitance of about 1 microFarad cm which means that to alter membrane voltage by 0.1 V, the membrane need only acquire or lose about 1 pmol of univalent ion cm of membrane. A univalent ion is one with a single positive (e.g. K) or negative (e.g. Cl) atomic charge. In a plant cell of about 650 pL, this represents a change in charge averaged over the entire cell volume of 12 nmol L !

The membrane voltage or membrane potential difference, as it is sometimes called, can be measured by inserting a fine capillary electrode into a plant cell. Membrane voltage is measured with respect to solution bathing the cell and in most plant cells the voltage is negative across the plasma membrane. That is, the cytoplasm has a charge of –0.1 to –0.3 V (–100 to –300 mV) at steady state with occasional transients that may give the membrane a positive voltage. The tonoplast membrane that surrounds the central vacuole is generally 20 to 40 mV more positive than the cytoplasm (still negative with respect to the outside medium).

Cell membrane voltages can be affected by ion pumps, diffusion potential and fixed charges on either side of the membrane.

Special mention needs to be made of one such fixed charge which arises from galacturonic acid residues in cell walls. Although cations move to neutralise this fixed negative change, there is still a net negative potential associated with cell walls (Donnan potential). In spite of being external to the plasma membrane, the Donnan potential is in series with it and probably adds to what we measure as the membrane potential with electrodes.

Most charge on macromolecules in the cytoplasm is also negative (e.g. nucleic acids, proteins) and because of their size it can be regarded as a fixed negative charge. This has osmotic consequences. Macromolecules and their balancing cations result in a relatively high osmotic pressure (up to 200 mmol L of total solutes) that results in inflow of water by osmosis and generation of P in most freshwater environments. In animal cells, where a cell wall is not present, body fluids must be regulated so that cell membranes do not rupture. For wall-less eukaryotes living in fresh water, a considerable amount of energy is expended in excreting water, for example by contractile vacuoles. Walled eukaryotes have the luxury of letting water enter the protoplasm by osmosis where P is developed rather than the cell expanding until it ruptures. Different ions have different permeabilities in membranes. Potassium, for example, is usually the most permeable ion, entering under most conditions about 10 to 100 times faster than Cl. Since ions diffuse at different rates across membranes, a slight charge imbalance occurs and gives rise to a membrane voltage. This voltage in turn slows down movement of the rapidly moving ion so that the counter-ion catches up. The result is that when net charge balance is achieved, a diffusion potential has developed that is a function of the permeabilities (P_{ion}) of all diffusible ions present and concentrations of each ion in each compartment. The Goldman equation describes this phenomenon and gives the membrane voltage (ΔE) that would develop due to diffusion of ions.

The superscripts refer to the inside (i) or outside (o) of the membrane and R, T, F and Care defined elsewhere. The concentration terms for Cl are reversed in the numerator and denominator compared to the cations. This is because Cl is the only anion represented. Many texts do not include H in the Goldman equation because, in spite of high permeability of H' diffusion of H is unlikely to have a strong effect on ΔE at such low (micromolar) concentrations.

However, membrane potential is occasionally dominated by the diffusion of H' indicating that H permeability must be exceedingly high. For example, local variations in pH cause alkaline bands to form on *Chara corallina* cells and in the leaves of aquatic plants at high pH.

THE NERNST EQUATION

When one ion has a very high permeability compared to all other ions in the system the membrane will behave as an ion-sensitive electrode for that ion. A pH electrode which is sensitive to H flux across a glass membrane serves as an analogy. In the case of a single ion, the Goldman equation can be reduced to the simpler Nernst equation that yields the equilibrium membrane potential which would develop for a particular concentration gradient across a membrane.

The Nernst equation is routinely used by electro-physiologists to calculate the equilibrium potential for each ion. Theoretical equilibrium potentials can then be compared with the actual membrane potential in order to decide whether the membrane is highly permeable to one particular ion. For example, in many plant cells there are K channels that open under particular circumstances. When this occurs, the membrane becomes highly permeable to K and the measured membrane potential very nearly equals the Nernst potential for K. The Nernst equation can also be used as a guide in deciding whether there is active transport through a membrane. For example, when the measured membrane potential is more negative than the most negative Nernst potential there must be active movement of charge across that membrane.

ACTIVE AND PASSIVE TRANSPORT

Plant cells acquire solutes and water across their membranes (plasma membrane and tonoplast) through the combined action of passive and active transport. During *passive transport,* a solute moves down its electrochemical potential gradient with no expenditure of energy. In *active transport* a solute moves against its electrochemical potential and hence energy input is required. The extra energy may be derived from the chemical energy released from hydrolysis of ATP or pyrophosphate (PP_i, a high-energy polymer of phosphate) or it may be derived from the movement of a cotransported solute or coupled solute down its electrochemical gradient. Coupling of downhill movement of one solute to uphill movement of another is a common feature of membrane transport.

Both the tonoplast and plasma membrane of plants and fungi contain pumps that move H across membranes. These pumps use chemical energy from the hydrolysis of ATP or PP_i. By moving positive charge out of the cytoplasm they establish a large membrane voltage (inside negative) and a steep pH gradient. The electrochemical gradient for H can be very large, the equivalent of about 400 mV. Bioenergetic analyses suggest that the energised proton pump in plasma membranes moves one H per ATP hydrolysed.

The H gradient established by the *primary active* H pumps is then used to drive coupled active movement of other solutes across membranes. These coupled transport systems can be referred to as *secondary active* transport to indicate that they rely on a previously established gradient of another ion. There are many examples of H -coupled transport in plants. One example is the uptake of a Cl ion, which is coupled to the influx of two H ions. Chloride must be actively transported across the plasma membrane because the membrane potential is usually so much more negative than the equilibrium potential for Cl. Coupling each Cl entering the cell to the inflow of two H ions means that there is a net charge transfer of +1 into the cell as each Cl enters. This is more energetically favourable than zero net charge and definitely more favourable than a net negative charge. Another example is H /sucrose cotransport which is important in the process of phloem loading.

Secondary active transport in plants is not exclusively driven by H. Na ion gradients sometimes drive secondary transport. This was first discovered in *Chara* but also occurs in some higher plants. A gene has been cloned from wheat roots for an Na -powered K transporter and when expressed in frog oocytes the transporter uses an inwardly directed Na gradient to drive uptake of K. Na -driven transport may have evolved in plants because under alkaline conditions (high external pH), the H gradient may not be sufficient to drive transport. Such conditions exist in seawater which is well buffered at a pH of about 8.3 but contains abundant Na. The discovery of Na -driven transport is especially interesting because as yet no one has identified a primary Na pump in plants. In animal cells, where Na gradients are generally used in secondary active transport, there is a primary Na pump that pumps three Na *out* and two K *in* for every ATP molecule hydrolysed. This pump does not seem to exist in plant cells but an Na –H exchanger (Na –H antiporter) has been shown to occur on the plasma membrane and tonoplast of some plants.

Solute movement across membranes by either active or passive processes can be regarded as analogous to an enzymic reaction. Transport proteins within membranes act as enzymes, catalysing solute transport by lowering the activation energy for transport. As with ordinary enzymes, the reaction may be coupled to the hydrolysis of ATP or some other high-energy molecule or it may proceed (energetically) downhill. The analogy with enzymes is especially useful to calculate affinity of the transport protein for its substrate. If we plot the rate of transport versus concentration of substrate being transported, a hyperbolic curve is often obtained. Analysis of this Michaelis–Menten curve reveals an affinity constant (K_m) and the maximum transport rate at saturating concentrations (equivalent to V_{max}).

The affinities of both passive and active transport vary widely between different transport systems in plants. Values of K_m can be in the range of micromoles per litre, or less, in which case we refer to the transport system as having a high affinity for its substrate. Alternatively, K_m may be tens of millimoles per litre in which case it would be referred to as a low-affinity transporter. There is no strict discrimination between active and passive transport systems on the basis of their affinity for substrates. Passive transporters can have high affinities and active transporters low affinities. Some transport systems do not saturate even if hundreds of millimoles of substrate per litre are present. This has often been attributed to transport through a pore in the membrane or a channel but there are many examples of channel-mediated transport that saturate at a few millimoles per litre and there is a Ca channel from plant plasma membrane that has a K_m in the micromolar range. Finally plants often have multiple transporters for the same solute, providing a range of affinities. This allows transport to proceed efficiently over a wide range of external concentrations, as would be encountered by a root in soil.

MEANS OF TRANSPORT

Diffusion: Movement by diffusion is passive, and may be from one part of the cell to the other, or from cell to cell, or over short distances, say, from the intercellular spaces of the leaf to the outside. No energy expenditure takes place. In diffusion, molecules move in a random fashion, the net result being substances moving from regions of higher concentration to regions of lower concentration. Diffusion is a slow process and is not dependent on a 'living system'. Diffusion is obvious in gases and liquids, but diffusion in solids rather than of solids is more likely. Diffusion is very important to plants since it is the only means for gaseous movement within the plant body. Diffusion rates are affected by the gradient of concentration, the permeability of the membrane separating them, temperature and pressure. Facilitated Diffusion:- A gradient must already be present for diffusion to occur. The diffusion rate depends on the size of the substances; obviously smaller substances diffuse faster. The diffusion of any substance across a membrane also depends on its solubility in lipids, the major constituent of the membrane. Substances soluble in lipids diffuse through the membrane faster. Substances that have a hydrophilic moiety, find it difficult to pass through the membrane; their movement has to be facilitated. Membrane proteins provide sites at which such molecules cross the membrane. They do not set up a concentration gradient: a concentration gradient must already be present for molecules to diffuse even if facilitated by the proteins. This process is called facilitated diffusion. In facilitated diffusion special proteins help move substances across membranes without expenditure of ATP energy. Facilitated diffusion cannot cause net transport of molecules from a low to a high concentration – this would require input of energy. Transport rate reaches a maximum when all of the protein transporters are being used (saturation). Facilitated diffusion is very specific: it allows cell to select substances for uptake. It is sensitive to inhibitors which react with protein side chains.

The proteins form channels in the membrane for molecules to pass through. Some channels are always open; others can be controlled. Some are large, allowing a variety of molecules to cross. The porins are proteins that form huge pores in the outer membranes of the plastids, mitochondria and some bacteria allowing molecules up to the size of small proteins to pass through molecule bound to the transport protein; the transport protein then rotates and releases the molecule inside the cell, e.g., water channels – made up of eight different types of aquaporins. Passive symports and antiports: Some carrier or transport proteins allow diffusion only if two types of molecules move together. In a symport, both molecules cross the membrane in the same direction; in an antiport, they move in opposite directions. When a molecule moves across a membrane independent of other molecules, the process is called uniport. Active Transport: Active transport uses energy to pump molecules against a concentration gradient. Active transport is carried out by membrane-proteins. Hence different proteins in the membrane play a major role in both active as well as passive transport. Pumps are proteins that use energy to carry substances across the cell membrane. These pumps can transport substances from a low concentration to a high concentration ('uphill' transport). Transport rate reaches a maximum when all the protein transporters are being used or are saturated. Like enzymes the carrier protein is very specific in what it carries across the membrane. These proteins are sensitive to inhibitors that react with protein side chains.

PLANT-WATER RELATIONS

Water is essential for all physiological activities of the plant and plays a very important role in all living organisms. It provides the medium in which most substances are dissolved. The protoplasm of the cells is nothing but water in which different molecules are dissolved and (several particles) suspended.

Distribution of water within a plant varies – woody parts have relatively very little water, while soft parts mostly contain water. Terrestrial plants take up huge amount water daily but most of it is lost to the air through evaporation from the leaves, i.e., transpiration.

Because of this high demand for water, it is not surprising that water is often the limiting factor for plant growth and productivity in both agricultural and natural environments.

WATER POTENTIAL:-

Water molecules possess kinetic energy. In liquid and gaseous form they are in random motion that is both rapid and constant. The greater the concentration of water in a system, the greater is its kinetic energy or 'water potential'. Hence, it is obvious that pure water will have the greatest water potential. If two systems containing water are in contact, random movement of water molecules will result in net movement of water molecules from the system with higher energy to the one with lower energy.

Thus water will move from the system containing water at higher water potential to the one having low water potential. This process of movement of substances down a gradient of free energy is called diffusion. Water potential is denoted by the Greek symbol Psi or ø and is expressed in pressure units such as pascals (Pa). By convention, the water potential of pure water at standard temperatures, which is not under any pressure, is taken to be zero.

SOLUTE POTENTIAL:-

If some solute is dissolved in pure water, the solution has fewer free water and the concentration of water decreases, reducing its water potential. Hence, all solutions have a lower water potential than pure water; the magnitude of this lowering due to dissolution of a solute is called solute potential or øs. øs is always negative. The more the solute molecules, the lower (more negative) is the $ø_s$. For a solution at atmospheric pressure

(water potential) $ø_w$ = (solute potential) $ø_s$.

PRESSURE POTENTIAL

If a pressure greater than atmospheric pressure is applied to pure water or a solution, its water potential increases. It is equivalent to pumping water from one place to another. Pressure can build up in a plant system when water enters a plant cell due to diffusion causing a pressure built up against the cell wall, it makes the cell turgid; this increases the pressure potential. Pressure potential is usually positive, though in plants negative potential or tension in the water column in the xylem plays a major role in water transport up a stem. Pressure potential is denoted as $ø_p$.

Water potential of a cell is affected by both solute and pressure potential. The relationship between them is as follows:

$$\Psi_w = \Psi_s + \Psi_p$$

OSMOSIS

Osmosis is the term used to refer specifically to the diffusion of water across a differentially- or semi-permeable membrane. Osmosis occurs spontaneously in response to a driving force. The net direction and rate of osmosis depends on both the pressure gradient and concentration gradient.

Water will move from its region of higher chemical potential (or concentration) to its region of lower chemical potential until equilibrium is reached. At equilibrium the two chambers should have the same water potential.

PLASMOLYSIS

If a plant cell is placed in a hypertonic solution, the plant cell loses water and hence turgor pressure, making the plant cell flaccid. Plants with cells in this condition wilt. Further water loss causes plasmolysis: pressure decreases to the point where the protoplasm of the cell peels away from the cell wall, leaving gaps between the cell wall and the membrane. Eventually cytorrhysis – the complete collapse of the cell wall – can occur. There are some mechanisms in plants to prevent excess water loss in the same way as excess water gain, but plasmolysis can be reversed if the cell is placed in a weaker solution (hypotonic solution). Stomata help keep water in the plant so it does not dry out. Wax also keeps water in the plant. The equivalent process in animal cells is called crenation.

The liquid content of the cell leaks out due to diffusion. The cell collapse and cell membrane pulls away from the cell wall(in plants). Most animal cells consist of only a phospholipid bilayer and not a cell wall, therefore shrinking up under such conditions. Plasmolysis only occurs in extreme conditions and rarely happens in nature. It is induced in the laboratory by immersing cells in strong saline or sugar solutions to cause exosmosis, often using Elodea plants or onion epidermal cells, which have coloured cell sap so that the process is clearly visible.

Plasmolysis can be of two types. It can be either concave plasmolysis or convex plasmolysis. Convex plasmolysis is always irreversible while concave plasmolysis is usually reversible.

IMBIBITION

Imbibition is a special type of diffusion when water is absorbed by solids – colloids – causing them to enormously increase in volume. The classical examples of imbibition are absorption of water by seeds and dry wood. The pressure that is produced by the swelling of wood had been used by prehistoric man to split rocks and boulders. If it were not for the pressure due to imbibition, seedlings would not have been able to emerge out of the soil into the open; they probably would not have been able to establish! Imbibition is also diffusion since water movement is along a concentration gradient; the seeds and other such materials have almost no water hence they absorb water easily. Water potential gradient between the absorbent and the liquid imbibed is essential for imbibition. In addition, for any substance to imbibe any liquid, affinity between the adsorbant and the liquid is also a pre-requisite.

LONG DISTANCE TRANSPORT OF WATER

The bulk movement of substances through the conducting or vascular tissues of plants is called translocation.

ABSORBPTION OF WATER

The responsibility of absorption of water and minerals is more specifically the function of the root hairs that are present in millions at the tips of the roots. Root hairs are thin-walled slender extensions of root epidermal cells that greatly increase the surface area for absorption. Water is absorbed along with mineral solutes, by the root hairs, purely by diffusion. Once water is absorbed by the root hairs, it can move deeper into root layers by two distinct pathways:

- *Apoplast pathway*
- *Symplast pathway*

Within a plant, the apoplast is the free diffusional space outside the plasma membrane. It is interrupted by the Casparian strip in roots, air spaces between plant cells and the cuticula of the plant. Structurally, the apoplast is formed by the continuum of cell walls of adjacent cells as well as the extracellular spaces, forming a tissue level compartment comparable to the symplast. The apoplastic route facilitates the transport of water and solutes across a tissue or organ. This process is known as apoplastic transport. The symplast of a plant is the inner side of the plasma membrane in which water (and low-molecular solutes) can freely diffuse.

The plasmodesmata allow the direct flow of small molecules such as sugars, amino acids, and ions between cells. Larger molecules, including transcription factors and plant viruses, can also be transported through with the help of actin structures. This allows direct cytoplasm to cytoplasm flow of water and other nutrients along concentration gradients. In particular, it is used in the root systems to bring in nutrients from soil. It moves these solutes from epidermis cells through the cortex into the endodermis and eventually the pericycle, where it can be moved into the xylem for long distance transport. It is contrasted with the apoplastic flow, which uses cell wall transport.

ROOT PRESSURE

As various ions from the soil are actively transported into the vascular tissues of the roots, water follows (its potential gradient) and increases the pressure inside the xylem. This positive pressure is called root pressure, and can be responsible for pushing up water to small heights in the stem.

Effects of root pressure is also observable at night and early morning when evaporation is low, and excess water collects in the form of droplets around special openings of veins near the tip of grass blades, and leaves of many herbaceous parts. Such water loss in its liquid phase is known as guttation.

Root pressure can, at best, only provide a modest push in the overall process of water transport. They obviously do not play a major role in water movement up tall trees. The greatest contribution of root pressure may be to re-establish the continuous chains of water molecules in the xylem which often break under the enormous tensions created by transpiration. Root pressure does not account for the majority of water transport; most plants meet their need by transpiratory pull.

TRANSPIRATION

Transpiration is the evaporative loss of water by plants. It occurs mainly through the stomata in the leaves. Besides the loss of water vapour in transpiration, exchange of oxygen and carbon dioxide in the leaf also occurs through pores called stomata (sing. : stoma). Normally stomata are open in the day time and close during the night. The immediate cause of the opening or closing of the stomata is a change in the turgidity of the guard cells. The inner wall of each guard cell, towards the pore or stomatal aperture, is thick and elastic. When turgidity increases within the two guard cells flanking each stomatal aperture or pore, the thin outer walls bulge out and force the inner walls into a crescent shape.

The opening of the stoma is also aided due to the orientation of the microfibrils in the cell walls of the guard cells. Cellulose microfibrils are oriented radially rather than longitudinally making it easier for the stoma to open. When the guard cells lose turgor, due to water loss (or water stress) the elastic inner walls regain their original shape, the guard cells become flaccid and the stoma closes.

Factors Affecting Transpiration:

- Temperature,
- Light,
- Humidity,
- Wind speed,
- Number and distribution of stomata,
- Number of stomatal aperture with guard cells stomata open,
- Water status of the plant,
- Canopy structure etc.

The transpiration driven ascent of xylem sap depends mainly on the following physical properties of water:

- Cohesion – mutual attraction between water molecules.

- Adhesion – attraction of water molecules to polar surfaces (such as the surface of tracheary elements).
- Surface Tension – water molecules are attracted to each other in the liquid phase more than to water in the gas phase.

These properties give water high tensile strength, i.e., an ability to resist a pulling force, and high capillarity, i.e., the ability to rise in thin tubes. In plants capillarity is aided by the small diameter of the tracheary elements – the tracheids and vessel elements. As water evaporates through the stomata, since the thin film of water over the cells is continuous, it results in pulling of water, molecule by molecule, into the leaf from the xylem. Also, because of lower concentration of water vapour in the atmosphere as compared to the substomatal cavity and intercellular spaces, water diffuses into the surrounding air. This creates a 'pull'.

TRANSPIRATION AND PHOTOSYNTHESIS

Transpiration has more than one purpose; it

- creates transpiration pull for absorption and transport of plants
- supplies water for photosynthesis
- transports minerals from the soil to all parts of the plant
- cools leaf surfaces, sometimes 10 to 15 degrees, by evaporative cooling
- maintains the shape and structure of the plants by keeping cells turgid

An actively photosynthesising plant has an insatiable need for water. Photosynthesis is limited by available water which can be swiftly depleted by transpiration. The humidity of rainforests is largely due to this vast cycling of water from root to leaf to atmosphere and back to the soil. The evolution of the C4 photosynthetic system is probably one of the strategies for maximising the availability of CO_2 while minimising water loss. C4 plants are twice as efficient as C3 plants in terms of fixing carbon (making sugar). However, a C4 plant loses only half as much water as a C3 plant for the same amount of CO_2 fixed.

UPTAKE AND TRANSPORT OF MINERAL NUTRIENTS

UPTAKE OF MINERAL IONS

Unlike water, all minerals cannot be passively absorbed by the roots. Two factors account for this:

(i) Minerals are present in the soil as charged particles (ions) which cannot move across cell membranes and
(ii) The concentration of minerals in the soil is usually lower than the concentration of minerals in the root.

Therefore, most minerals must enter the root by active absorption into the cytoplasm of epidermal cells. This needs energy in the form of ATP. The active uptake of ions is partly responsible for the water potential gradient in roots, and therefore for the uptake of water by osmosis. Some ions also move into the epidermal cells passively. Ions are absorbed from the soil by both passive and active transport. Specific proteins in the membranes of root hair cells actively pump ions from the soil into the cytoplasms of the epidermal cells. Like all cells, the endodermal cells have many transport proteins embedded in their plasma membrane; they let some solutes cross the membrane, but not others.

TRANSLOCATION OF MINERAL IONS

After the ions have reached xylem through active or passive uptake, or a combination of the two, their further transport up the stem to all parts of the plant is through the transpiration stream. The chief sinks for the mineral elements are the growing regions of the plant, such as the apical and lateral meristems, young leaves, developing flowers, fruits and seeds, and the storage organs. Unloading of mineral ions occurs at the fine vein endings through diffusion and active uptake by these cells. Mineral ions are frequently remobilised, particularly from older, senescing parts. Older dying leaves export much of their mineral content to younger leaves. Similarly, before leaf fall in decidous plants, minerals are removed to other parts. Elements most readily mobilised are phosphorus, sulphur, nitrogen and potassium. Some elements that are structural components like calcium are not remobilised. An analysis of the xylem exudates shows that though some of the nitrogen travels as inorganic ions, much of it is carried in the organic form as amino acids and related compounds. Similarly, small amounts of P and S are carried as organic compounds. In addition, small amount of exchange of materials does take place between xylem and phloem.

PHLOEM TRANSPORT

Food, primarily sucrose, is transported by the vascular tissue phloem from a source to a sink. Usually the source is understood to be that part of the plant which synthesises the food, i.e., the leaf, and sink, the part that needs or stores the food. But, the source and sink may be reversed depending on the season, or the plant's needs. Sugar stored in roots may be mobilised to become a source of food in the early spring when the buds of trees, act as sink; they need energy for growth and development of the photosynthetic apparatus. Since the source-sink relationship is variable, the direction of movement in the phloem can be upwards or downwards, i.e., bi-directional. This contrasts with that of the xylem where the movement is always unidirectional, i.e., upwards. Hence, unlike one-way flow of water in transpiration, food in phloem sap can be transported in any required direction so long as there is a source of sugar and a sink able to use, store or remove the sugar. Phloem sap is mainly water and sucrose, but other sugars, hormones and amino acids are also transported or translocated through phloem.

THE PRESSURE FLOW OR MASS FLOW HYPOTHESIS

The accepted mechanism used for the translocation of sugars from source to sink is called the pressure flow hypothesis. As glucose is prepared at the source (by photosynthesis) it is converted to sucrose (a dissacharide). The sugar is then moved in the form of sucrose into the companion cells and then into the living phloem sieve tube cells by active transport. This process of loading at the source produces a hypertonic condition in the phloem. Water in the adjacent xylem moves into the phloem by osmosis. As osmotic pressure builds up the phloem sap will move to areas of lower pressure. At the sink osmotic pressure must be reduced. Again active transport is necessary to move the sucrose out of the phloem sap and into the cells which will use the sugar – converting it into energy, starch, or cellulose. As sugars are removed, the osmotic pressure decreases and water moves out of the phloem.

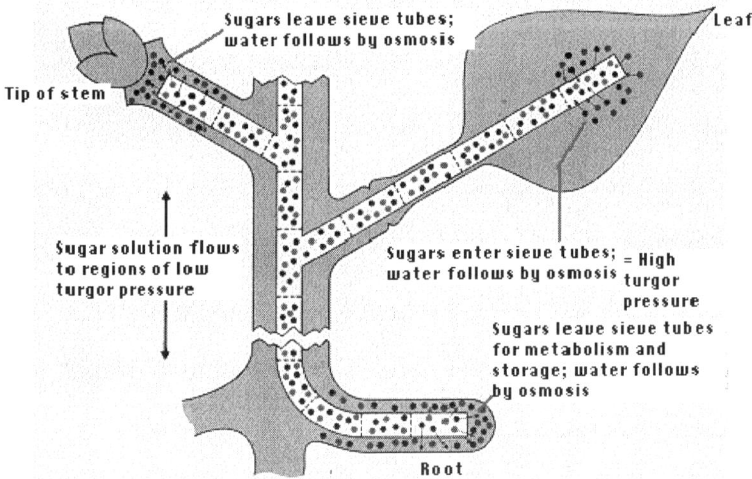

To summarise, the movement of sugars in the phloem begins at the source, where sugars are loaded (actively transported) into a sieve tube. Loading of the phloem sets up a water potential gradient that facilitates the mass movement in the phloem.

Phloem tissue is composed of sieve tube cells, which form long columns with holes in their end walls called sieve plates. Cytoplasmic strands pass through the holes in the sieve plates, so forming continuous filaments. As hydrostatic pressure in the phloem sieve tube increases, pressure flow begins, and the sap moves through the phloem. Meanwhile, at the sink, incoming sugars are actively transported out of the phloem and removed Sugars leave sieve tube for metabolism and storage; water follows by osmosis as complex carbohydrates. The loss of solute produces a high water potential in the phloem, and water passes out, returning eventually to xylem.

2

Cell Membranes and Transport

MEMBRANES

All living cells have something known as a cell membrane. This selectively-permeable membrane controls the exchange of materials, receives hormone messages and is very thin. It can be described as a phospholipid bi-layer - meaning that it is made from phospholipid molecules and has two layers.

STRUCTURE

The phospholipid bi-layer is so thin it can barely even be seen by an electron microscope - a x100,000 magnification is required, and only shows a double black line around 7 nm wide. Since we cannot properly see the membrane, we have to take what we know about it and create a model - in this case known as a fluid mosaic model.

FLUID MOSAIC MODEL

A diagram of the fluid mosaic model can be seen below.

Features of the fluid mosaic model;

- The membrane primarily consists of a bilayer of phospholipid molecules. These molecules can move about by diffusion in their own layer.
- Width is about 7 nm on average
- Some of the phospholipids are saturated and some are unsaturated. This affects the fluidity of the membrane; an unsaturated membrane means a more fluid membrane. This is due to the kink in unsaturated tails causing the molecules to not sit closely together.
- Phospholipid tails point inwards, facing each other, meaning that inside the membrane it is non-polar hydrophobic.
- The protein molecules within the structure can move around although some are fixed to structures inside the cell and do not move. Also, some of them span the width of the membrane, some are only on the inner layer and some on the outer layer.

- Many proteins and lipids have short carbohydrate chains attached to them, forming glycoproteins and glycolipids.
- Contains cholesterol.

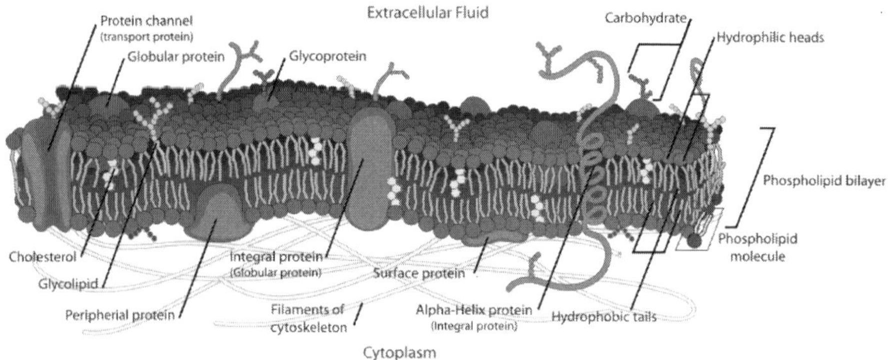

COMPONENTS

About several components that may be new to you, their structures and roles are below.

PHOSPHOLIPIDS

Phospholipids have hydrophilic heads and hydrophobic tails, the importance of which should be becoming clear. A phospholipid bi-layer forms the body of the membrane, creating a hydrophobic interior and isolating the cell from the outside environment. Do not forget that organelles within a cell are also often surrounded by a membrane, and this too is a phospholipid membrane.

PROTEINS

Proteins have a variety of functions within membranes. Some membrane proteins are enzymes, catalysing reactions such as the ones in the surface of the cell membrane. However, most are transport proteins, providing hydrophilic channels through the bi-layer for ions and polar molecules to pass through. They are usually specific to a certain ion or molecule. They can control the types of substances that can leave or enter the cell. The membranes of organelles also hold proteins. Can you think of any EG.'s?

GLYCOPROTEINS AND GLYCOLIPIDS

Lipid and proteins on the cell membrane surface often have short carbohydrate chains protruding out from the cell surface, known as glycolipids and glycoproteins. They form hydrogen bonds with the water molecules surrounding the cell and thus help to stabilise membrane structure. However, more importantly, they are used as receptor molecules, binding with hormones or neurotransmitters to trigger a series of chemical reactions within the cell itself. Using insulin as an example, only some cells within the body (liver, muscles), have receptors for insulin and as such, insulin can be released to the entire body without upsetting anything as any cell without an insulin receptor will not be affected. In this way they make the cell specific.

CHOLESTEROL

Cholesterol helps to regulate fluidity of the membrane and also to provide mechanical stability of the membranes - without it cells will burst open as their membranes break. Their hydrophobic regions help to prevent ions or polar molecules inadvertently passing through the membrane.

TRANSPORT

Transport across the phospholipid bi-layer is regulated, and it is an effective barrier - but exchange is necessary.

DIFFUSION

Diffusion is defined as the net motion of a substance from an area of high concentration to an area of low concentration, 'down a concentration gradient'.

Factors which affect diffusion;

- How steep the concentration gradient is.
- Temperature.
- Surface area.
- Type of molecule.

Molecules and ions that are small enough can cross membranes easily, regardless of polarity, but large polar molecules such as glucose cannot diffuse through a cell membrane. They can only pass through hydrophilic protein channels - this process is known as facilitated diffusion. All the factors that affect diffusion affect facilitated diffusion, and an additional one - how many transport proteins are available.

Facilitated diffusion:

OSMOSIS

Osmosis is best described as a special type of diffusion involving water molecules (H_2O) only.

See these two diagrams:

The first diagram is the experiment just set up, the 'before diagram', and as you can see, water is moving from the more dilute substance to the less dilute solution, separated by a partially permeable membrane - down the concentration gradient. The second picture shows the substance after it has been left for a while - there is the same concentration of solute molecules and thus the same concentration of water molecules, this is known as equilibrium. The fact that this is the movement of water molecules alone that has bought about the equilibrium is characteristic of osmosis.

WATER POTENTIAL

The propensity for water molecules to move from one place to another is known as the water potential, and the symbol for water potential is the Greek letter 'psi', Ø. Water always moves from a area of high water potential to a low water potential area - equilibrium as mentioned before is the equaling of two adjacent (separated by a selectively-permeable membrane) water potentials. Pure water has a water potential of 0, and any solutes make the water potential negative, the degree to which they do called the solute potential.

PRESSURE POTENTIAL

Net movement of water from a high to low concentration gradient can be prevented or slowed by increasing the pressure in the low concentration gradient solution, since pressure increases the water potential.

IN PLANT CELLS

Plants have cell walls, and thus pressure potential is especially important - if the volume of the cell increases as a result of water entering the cell via osmosis, the cell starts to push against the cell wall and pressure builds up rapidly. This pressure increases the water potential of the cell so that water stops entering it, creating a 'false-equilibrium' (in that pressure helps it) preventing it from bursting. When a cell is fully inflated it is described as turgid.

The opposite to turgidity is plasmolysis - when a cell is placed in a concentrated sucrose solution, the cell begins to shrink away from the cell wall, creating a pressure potential of 0, and so the water potential is equal to its solute potential. Eventually cytorrhysis – the complete collapse of the cell wall – can occur. There is no mechanism in plants to prevent excess water loss in the same way as excess water gain, but plasmolysis can be reversed if the cell is placed in a weaker solution. The equivalent process in animal cells is called crenation. The liquid content of the cell leaks out due to diffusion. Plasmolysis is extremely rare in nature.

IN ANIMAL CELLS

In animal cells however, there is no cell wall, and so if the water potential of the solution around the cell is too high, the cell will swell and burst, but if it is too low, the cell shrivels or shrinks. This is why it s important to keep a constant water potential inside animal bodies.

ACTIVE TRANSPORT

Active transport is defined as the energy-consuming transport of molecules or ions across a membrane against a concentration gradient, made possible by transferring energy from respiration. The energy is supplied by

ATP, and is used to make the transport protein change its 3-d shape, transferring the molecules or ions across the membrane in the process. Something to note is that cells that do a lot of active transport are likely to have many mitochondrion to provide the energy for it.

It is particularly important in reabsoprtion in the kidneys where certain useful molecules must be reabsorbed into the blood after filtration. In plants it is used to load sugar from photosynthesising cells into the phloem tissue for transport. Facilitated Diffusion of Ions

Facilitated diffusion of ions takes place through proteins, or assemblies of proteins, embedded in the plasma membrane. These transmembrane proteins form a water-filled channel through which the ion can pass down its concentration gradient.

The transmembrane channels that permit facilitated diffusion can be opened or closed. They are said to be "gated". Some types of gated ion channels:

- Ligand-gated
- Mechanically-gated
- Voltage-gated
- Light-gated

Ligand-gated ion channels. Many ion channels open or close in response to binding a small signaling molecule or "ligand". Some ion channels are gated by extracellular ligands; some by intracellular ligands. In both cases, the ligand is not the substance that is transported when the channel opens. External ligands

External ligands (shown here in green) bind to a site on the extracellular side of the channel.

Examples:

- Acetylcholine (ACh). The binding of the neurotransmitter acetylcholine at certain synapses opens channels that admit Na+ and initiate a nerve impulse or muscle contraction.
- Gamma amino butyric acid (GABA). Binding of GABA at certain synapses — designated GABAA — in the central nervous system admits Cl- ions into the cell and inhibits the creation of a nerve impulse. [More]

Internal ligands

Internal ligands bind to a site on the channel protein exposed to the cytosol.

Examples:

- "Second messengers", like cyclic AMP (cAMP) and cyclic GMP (cGMP), regulate channels involved in the initiation of impulses in neurons responding to odors and light respectively.
- ATP is needed to open the channel that allows chloride (Cl-) and bicarbonate (HCO3-) ions out of the cell. This channel is defective in patients with cystic fibrosis. Although the energy liberated by

the hydrolysis of ATP is needed to open the channel, this is not an example of active transport; the ions diffuse through the open channel following their concentration gradient.

MECHANICALLY-GATED ION CHANNELS

Examples:

- Sound waves bending the cilia-like projections on the hair cells of the inner ear open up ion channels leading to the creation of nerve impulses that the brain interprets as sound. [More]
- Mechanical deformation of the cells of stretch receptors opens ion channels leading to the creation of nerve impulses. [More]

VOLTAGE-GATED ION CHANNELS

In so-called "excitable" cells like neurons and muscle cells, some channels open or close in response to changes in the charge (measured in volts) across the plasma membrane.

Example: As an impulse passes down a neuron, the reduction in the voltage opens sodium channels in the adjacent portion of the membrane. This allows the influx of Na+ into the neuron and thus the continuation of the nerve impulse.

Some 7000 sodium ions pass through each channel during the brief period (about 1 millisecond) that it remains open. This was learned by use of the patch clamp technique:

- A very fine pipette (with an opening of about 0.5 μm) is pressed against the plasma membrane of
- Either an intact cell or
- The plasma membrane can be pulled away from the cell and the preparation placed in a test solution of desired composition.
- Current flow through a single ion channel can then be measured.

Such measurements reveal that each channel is either fully open or fully closed; that is, facilitated diffusion through a single channel is "all-or-none".

This technique has provided so much valuable information about ion channels that its inventors, Erwin Neher and Bert Sakmann, were awarded a Nobel Prize in 1991. Facilitated Diffusion of Molecules

Some small, hydrophilic organic molecules, like sugars, can pass through cell membranes by facilitated diffusion.

Once again, the process requires transmembrane proteins. In some cases, these — like ion channels — form water-filled pores that enable the molecule to pass in (or out) of the membrane following its concentration gradient.

Example: Maltoporin. This homotrimer in the outer membrane of E. coli forms pores that allow the disaccharide maltose and a few related molecules to diffuse into the cell.

Another example: The plasma membrane of human red blood cells contain transmembrane proteins that permit the diffusion of glucose from the blood into the cell. Note that in all cases of facilitated diffusion through channels, the channels are selective; that is, the structure of the protein admits only certain types of molecules through.

Whether all cases of facilitated diffusion of small molecules use channels is yet to be proven. Perhaps some molecules are passed through the membrane by a conformational change in the shape of the transmembrane protein when it binds the molecule to be transported. In either case, the interaction between the molecule being transported and its transporter resembles in many ways the interaction between an enzyme and its substrate. Link to a discussion of the energetics of enzyme-substrate interactions. Active Transport

Active transport is the pumping of molecules or ions through a membrane against their concentration gradient. It requires:

- A transmembrane protein (usually a complex of them) called a transporter and
- Energy. The source of this energy is ATP.

The energy of ATP may be used directly or indirectly.

- Direct Active Transport. Some transporters bind ATP directly and use the energy of its hydrolysis to drive active transport.
- Indirect Active Transport. Other transporters use the energy already stored in the gradient of a directly-pumped ion. Direct active transport of the ion establishes a concentration gradient. When this is relieved by facilitated diffusion, the energy released can be harnessed to the pumping of some other ion or molecule.

Link to a quantitative analysis of these processes. Direct Active Transport

1. The Na^+/K^+ ATPase

The cytosol of animal cells contains a concentration of potassium ions (K+) as much as 20 times higher than that in the extracellular fluid. Conversely, the extracellular fluid contains a concentration of sodium ions (Na+) as much as 10 times greater than that within the cell.

These concentration gradients are established by the active transport of both ions. And, in fact, the same transporter, called the Na+/K+ ATPase, does both jobs. It uses the energy from the hydrolysis of ATP to

- Actively transport 3 Na+ ions out of the cell
- For each 2 K+ ions pumped into the cell.

This accomplishes several vital functions:

- It helps establish a net charge across the plasma membrane with the interior of the cell being negatively charged with respect to the exterior. This resting potential prepares nerve and muscle cells for the propagation of action potentials leading to nerve impulses and muscle contraction.
- The accumulation of sodium ions outside of the cell draws water out of the cell and thus enables it to maintain osmotic balance (otherwise it would swell and burst from the inward diffusion of water).

- The gradient of sodium ions is harnessed to provide the energy to run several types of indirect pumps.

The crucial roles of the Na+/K+ ATPase are reflected in the fact that almost one-third of all the energy generated by the mitochondria in animal cells is used just to run this pump. 2. The H+/K+ ATPase

The parietal cells of your stomach use this pump to secrete gastric juice. These cells transport protons (H+) from a concentration of about 4 x 10-8 M within the cell to a concentration of about 0.15 M in the gastric juice (giving it a pH close to 1). Small wonder that parietal cells are stuffed with mitochondria and uses huge amounts of energy as they carry out this three-million fold concentration of protons. 3. The Ca2+ ATPases

A Ca2+ ATPase is located in the plasma membrane of all eukaryotic cells. It uses the energy provided by one molecule of ATP to pump one Ca2+ ion out of the cell. The activity of these pumps helps to maintain the ~20,000-fold concentration gradient of Ca2+ between the cytosol (~ 100 nM) and the ECF (~ 20 mM). [More] In resting skeletal muscle, there is a much higher concentration of calcium ions (Ca2+) in the sarcoplasmic reticulum than in the cytosol. Activation of the muscle fiber allows some of this Ca2+ to pass by facilitated diffusion into the cytosol where it triggers contraction. [Link to discussion]. After contraction, this Ca2+ is pumped back into the sarcoplasmic reticulum. This is done by another Ca2+ ATPase that uses the energy from each molecule of ATP to pump 2 Ca2+ ions.

Pumps 1. - 3. are designated P-type ion transporters because they use the same basic mechanism: a conformational change in the proteins as they are reversibly phosphorylated by ATP. And all three pumps can be made to run backward. That is, if the pumped ions are allowed to diffuse back through the membrane complex, ATP can be synthesized from ADP and inorganic phosphate. 4. ABC Transporters ABC ("ATP-Binding Cassette") transporters are transmembrane proteins that

- Expose a ligand-binding domain at one surface and a
- ATP-binding domain at the other surface.

The ligand-binding domain is usually restricted to a single type of molecule, so this is called the phosphilid layer. The ATP bound to its domain provides the energy to pump the ligand across the membrane.

The human genome contains 48 genes for ABC transporters. Some examples:

- CFTR — the cystic fibrosis transmembrane conductance regulator
- TAP, the transporter associated with antigen processing. [Discussion]
- The transporter that liver cells use to pump the salts of bile acids out into the bile.
- ABC transporters that pump chemotherapeutic drugs out of cancer cells thus reducing their effectiveness.

ABC transporters must have evolved early in the history of life. The ATP-binding domains in archaea, eubacteria, and eukaryotes all share a homologous structure, the ATP-binding "cassette". Indirect Active Transport Indirect active transport uses the downhill flow of an ion to pump some other molecule or ion against its gradient. The driving ion is usually sodium (Na+) with its gradient established by the Na+/K+ ATPase. Symport Pumps

In this type of indirect active transport, the driving ion (Na+) and the pumped molecule pass through the membrane pump in the same direction. Examples:

- The Na+/glucose transporter. This transmembrane protein allows sodium ions and glucose to enter the cell together. The sodium ions flow down their concentration gradient while the glucose molecules are pumped up theirs. Later the sodium is pumped back out of the cell by the Na^+/K^+ ATPase.

  External Link

  The Na^+/glucose transporter is used to actively transport glucose out of the intestine and also out of the kidney tubules and back into the blood.

  The energy relationships for these processes can be quantified. Link to a discussion.
- All the amino acids can be actively transported, for example
  - Out of the kidney tubules and into the blood [Example]
  - The reuptake of Glu from the synapse back into the presynaptic neuron by sodium-driven symport pumps.
- The Na^+/iodide transporter. This symporter pumps iodide ions into the cells of the thyroid gland (for the manufacture of thyroxine) and also into the cells of the mammary gland (to supply the baby's need for iodide).
- The permease encoded by the lac operon of E. coli that transports lactose into the cell.

ANTIPORT PUMPS

In antiport pumps, the driving ion (again, usually sodium) diffuses through the pump in one direction providing the energy for the active transport of some other molecule or ion in the opposite direction.

Example: Ca2+ ions are pumped out of cells by sodium-driven antiport pumps [Link]. Antiport pumps in the vacuole of some plants harness the outward facilitated diffusion of protons (themselves pumped into the vacuole by a H+ ATPase)

- To the active inward transport of sodium ions. This sodium/proton antiport pump enables the plant to sequester sodium ions in its vacuole. Transgenic tomato plants that overexpress this sodium/ proton antiport pump are able to thrive in saline soils too salty for conventional tomatoes.
- To the active inward transport of nitrate ions (NO3").

BULK TRANSPORT

Bulk transport can be defined as the movement of large quantities of materials into or out of cells, endocytosis and exocytosis, respectively.

Exocytosis is the process by which materials are removed from cells - for example the secretion of digestive enzymes, where vesicles from the Golgi apparatus carry the enzymes to the cell surface, bind to it, and release their contents. See: Endocytosis is the reverse of exocytosis and involves the engulfing of the material by the cell to form a small sac inside the cell. The most common form is phagocytosis, performed by phagocytes - an example of which would be white blood cells engulfing bacteria. The second form of endocytosis is pinocytosis, the bulk uptake of liquid, and the human egg cell takes up nutrients from cells that surround it by this method.

EXCHANGE SURFACES

This relates to how substances move in and out of whole organisms, using two examples.

MAMMALIAN LUNGS

All mammals need a supply of oxygen to use in respiration - and in mammals the cells that require the oxygen are too far for diffusion to be effective and so they have a specialised gaseous exchange surface, where oxygen can diffuse into the body and carbon dioxide can diffuse out. In humans, this exchange surface is the alveoli in the lungs, and as you can see in the diagram below, each alveolus is tiny but collectively they have a huge surface area, totalling around 70m in an adult. This increases the net rate of diffusion. Alveoli also have extremely thin walls, no more than 0.5 μm thick, directly next to blood capillaries also with very thin cell membranes. This thinness allows diffusion to be extremely speedy.

Since diffusion only works down a concentration gradient, and the steeper said concentration gradient, the faster diffusion, the concentration gradient between the alveoli and the blood must be kept steep constantly to ensure speedy diffusion. Carbon dioxide is rapidly breathed out, meaning that carbon dioxide transported to the alveoli will be quickly transported into them and breathed out, and deoxygenated blood is constantly kept coming to the alveoli, meaning that oxygen diffuses into it quickly.

PLANT ROOTS

Root hairs of a flowering plant are also specialised exchange surfaces - they are extensions of the cells that make up the epidermis of a root, and are each around 200-250μm across, providing an enormous total surface area that is in contact with the soil surrounding the root. Both water and mineral ions are absorbed by diffusion and facilitated diffusion/active transport respectively.

MULTICELLULAR PLANTS

WATER TRANSPORT

Water enters a plant through the hairs on the root, and moves across the root cells into the xylem, which transports it up and around the plant. That, and solutes are moved around by the xylem and the phloem, using the root, stem and plant.

Root

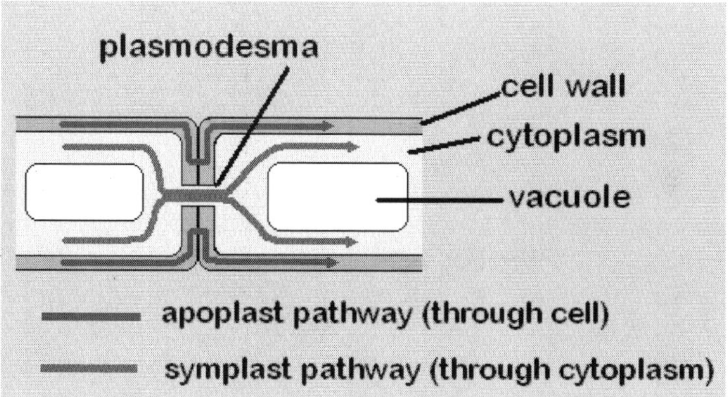

Fig. The Apoplast and Symplast Pathways

Water enters the root through the root hairs, and then takes one of two paths (apoplast and symplast) to the xylem vessel.

SOIL TO ROOT HAIR

A root hair is a simple extension of the epidermis of a root cell, and reach into the soil to absorb water. It exists to increase the surface area and therefore the rate at which water can be absorbed. Some plants have fungi which act like fine roots, absorbing nutrients from the soil for the plant. Water moves into the root hair cells because it is moving down a water potential gradient, since a root cell has a relatively low water potential due to its inorganic ions and organic substances. Water will enter through the membrane and into the cytoplasm and vacuole.

ROOT HAIR TO XYLEM

From the root hair cells, water again moves down a concentration gradient toward the xylem, and can take one of three paths - apoplast and symplast and vacoular.

The apoplast pathway is where water takes a route going from cell wall to cell wall, not entering the cytoplasm at any point. The symplast pathway is where water moves between cytoplasm/vacuoles of adjacent cells. However, the apoplast pathway can only take water a certain way, near the xylem, the Casparian strip forms an impenetrable barrier to water in the cell walls, and it must move into the cytoplasm to continue. This gives the plant control over the ions that enter its xylem vessels, since water must cross a plasma membrane to get there. The vacoular pathway moves molecules through the vacuoles only of the plant.

XYLEM

The xylem is constructed of three main elements;

- Vessel elements, including tracheids - cells involved in water transport
- Fibres - elongated cells with lignified walls that support the plant
- Parenchyma cells - normal plant cells, except no chloroplasts.

XYLEM VESSELS

These vessel elements make up the xylem - and are many elongated cells laid end to end, and are normal plant cells with their walls strengthened by lignin, a hard strong substance that is impermeable to water, and is designed to provide structure and strength to the plant. When these plant cells are strengthened by lignin, the cell inside dies, leaving a space inside. However, in some plasmodesmata, there was no lignin laid down and these appear as gaps in the xylem vessel, know as pits. These have permeable unthickened cellulose cell wall.

Thus, a continuous tube is formed, known as the xylem vessel. Xylem vessels are huge

TRACHEIDS

Tracheids are dead cells with lignified walls, but they do not have open ends and thus do not form vessels - their ends are tapered. All plants have them, but primitive plants use them as main conducing tissue.

TRANSLOCATION

The transport of soluble organic substances (sometimes called assimilates) within a plant is known as translocation. The solutes are transported in sieve elements, found in the phloem tissues, along with other as companion cells, parenchyma and fibres.

COMPONENTS

SIEVE ELEMENTS

Sieve Elements are specialised cells, with few mitochondrion and endoplasmic reticulum, no nucleus or ribosomes. Where two ends of sieve elements meet, a sieve plate is formed, made from walls of both elements, with large pores allowing free flow of liquids between them.

COMPANION CELLS

Companion cells are normal plant cells with high numbers of mitochondria and ribosomes, and have many plasmodesmata pass through to their neighbouring sieve cell walls, making direct contact between the cytoplasms of companion cell and sieve element.

MASS FLOW

Mass flow is the theory by which we think solute transport occurs in plants. Any area where sucrose is produced in a plant is known as a *source*, and any area where it is taken out (usually, used in respiration) is known as a *sink*.

- Sucrose is actively transported into the sieve tubes of the phloem at the source (i.e. source), lowering the water potential inside the sieve and so water enters the tubes via osmosis, creating a higher pressure inside the sieve tubes at the source.
- At the sink, sugars leave the phloem to be used up, increasing the water potential inside the sieve tubes, so water leaves via osmosis, lowering the pressure inside the sieve tubes.
- The result is a pressure gradient between source to sink, pushing sugars to where they're needed.

EVIDENCE

Supporting evidence

- When the phloem is cut, sap oozes out, showing a pressure gradient.
- Suitable water potential gradient between leaves and other plants, in theory.
- Phloem sap has a high pH, which is to be expected since hydrogen ions are actively transported out of the cell.
- ATP is present in phloem sieve elements in high numbers since it is required for active transport of hydrogen ions.

Evidence against

- Sugar travels to many different sinks.
- Sieve plates are a barrier to mass flow.
- Doesn't require living cells, but phloem cells are alive.

TRANSPIRATION

Transpiration is the loss of water vapour from the leaves of a plant.

XYLEM TO LEAF

As water evaporates from the leaf, a constantly occurring process, more water is taken in to replace it. The removal of water reduces the hydrostatic pressure (pressure exerted by a liquid). Since this pressure becomes lower at the top of the xylem vessel than at the bottom, this pressure difference causes water to move up the xylem vessels, just as in a straw. This process is known as mass flow - as long with the fact that water molecules move together as a body of water - aided by water's property of being cohesive, and attracted to the lignin in the walls of the xylem vessels, known as adhesion.

Once water is in the leaf, it can be lost through the stomata, if there is a concentration gradient that it can go down, which are small pores in direct contract with the air outside. This process is known as transpiration.

ROOT PRESSURE

Plants can also increase the hydrostatic pressure at the bottom of the vessels, changing the pressure difference. They do this by cells surrounding the xylem vessels to use active transport to pump solutes across their membranes and into the xylem, lowering the water potential of the solution in the xylem, thus drawing in water from the surrounding root cells. The influx of water at the bottom of the xylem increases the pressure.

TRANSPIRATION RATE

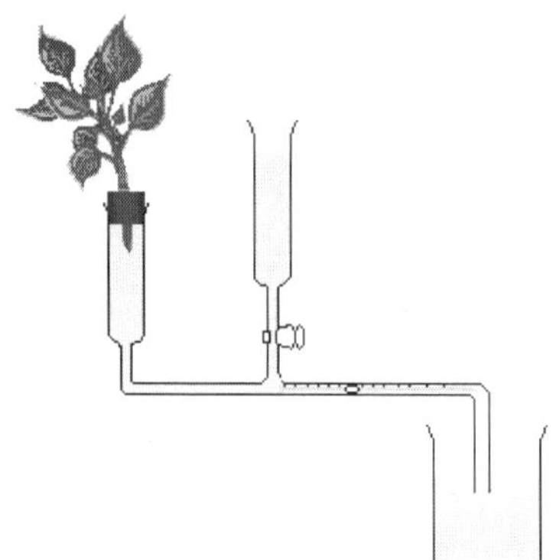

Fig. A potometer

The rate of transpiration is affected by:

- Humidity, which affects the water potential gradient outside the leaf, by making the gradient more or less steep. Low levels of humidity would cause the gradient to be steep, meaning that transpiration would increase as water molecules would find it easier to escape the leaves. A high humidity level would have the opposite effect, reducing the transpiration rate.
- Wind speed, which affects the water potential gradient outside the leaf, the faster the speed, the less humid the area around the leaf becomes, as a result, the water potential gradient would be come more steep and more evapotranspiration would occur.
- Temperature, which makes it easier for water molecules to escape

from the leaves (this happens when it is hot). as temperature increases, the water molecules are able to move more rapidly. The evaporation rate also increases with temperature. Warm air is also able to hold more moisture than cold air, meaning that water can diffuse into warm air more easily, and this contributes to the transpiration rate.

- Light, Light stimulates the stomata to open allowing gas exchange for photosynthesis, and as a side effect this also increases transpiration. This is a problem for some plants as they may lose water during the day and wilt.

The transpiration rate is measured by a potometer, a device that measures the rate at which a plant stem takes up water, since measuring transpiration directly is difficult.

XEROPHYTES

Xerophytes are plants which have various adaptions (below) to reduce transpiration and thus water loss.

- Reduced stomata
- Swollen stem, storing water
- Swollen leaves to store water
- Huge surface area for roots
- Leaves reduced to spines (cactus), reducing surface area for water loss
- Take in CO_2 at night so they can close their stomata in the day to reduce water loss
- Thick, waxy water-resistant cuticle on the epidermis to reduce water loss
- Sunken stomata (in pits), where water vapour is sheltered from the wind
- Hairlike projections, called trichomes, on the leaf surface to reduce the effect of high wind speed

IDENTIFICATION OF GENES INVOLVED IN METAL TRANSPORT IN PLANTS

Plants obtain mineral nutrients from the soil. If they are growing in soil with high levels of metals, they will take up an excess of what is needed for growth. Depending on the species, this can be detrimental to growth—or lethal—and can greatly limit the growth range of plants and the productivity of agricultural species. However, some plants have adapted to living in soil containing excess metals. A portion of these species will even hyperaccumulate the metals, leading to exceedingly high levels of metals in the plant tissues. The metal most commonly accumulated is nickel (Ni).

How and why certain plants are able to accumulate—and tolerate—high levels of potentially toxic compounds has spawned diverse areas of research, including an article by Talke et al., "Zinc-dependent global transcriptional control, transcriptional deregulation, and higher gene copy number for genes in metal homeostasis of the hyperaccumulator *Arabidopsis halleri*," which appeared in the September 2006 issue of *Plant Physiology*.

BACKGROUND

Interestingly, most of Ni and zinc (Zn) hyperaccumulators belongs to one family, the Brassicaceae, which includes the well-studied *Thlaspi caerulescens* and*Arabidopsis halleri*. Hyperaccumulators such as these typically accumulate the metals in the aboveground biomass through bulk flow of the metals in the xylem from root to shoot. Prior to this, the metals must first be translocated from the root symplast into the xylem apoplast, and in most instances the transporter proteins involved in this process have not been identified. This is a saturable process limited not only by the number of transport proteins present, but also by the variation in the transporters with respect to transport rate, substrate affinity, and substrate specificity. Another transport step occurs from the xylem into the leaf cells. At the tissue level, metal may be accumulated in the epidermis and trichomes, while at the cellular level, these excess metals are typically accumulated in the vacuole or cell wall. The metals are often bound by chelators, which are believed to play a role in detoxification of the metals.

The ability of plants to remove organic contaminants such as metals and accumulate them in aboveground biomass has been taken advantage of in the remediation of contaminated soils. However, not all metal-hyperaccumulating plants have high biomass, a "requirement" for successful use of a plant for phytoremediation. Thus, an important question is what confers the ability to tolerate (elevated) levels of a metal that would be lethal or seriously inhibit the growth of a closely related species, or, more specifically, what makes a plant a hyperaccumulator?

WHAT WAS SHOWN

A. halleri is a member of the sister clade to *Arabidopsis thaliana*, but unlike *A. thaliana* it is a metal hyperaccumulator of Zn and cadmium (Cd). As is typical for hyperaccumulators, the Zn hyperaccumulation in *A. halleri* is partly due to increased partitioning of the metal from roots to shoots. Using cross-species transcript profiling, Talke et al. (2006) identified a set of candidate genes that are more highly expressed in *A. halleri* than in *A. thaliana* when grown under elevated levels of Zn and a set in *A. halleri* that is transcriptionally regulated by Zn. The change in transcript level of a subset of these candidate genes was further analyzed using real-time PCR and 29 of these were found to encode putative metal homeostasis proteins. Although 11 of these 29 were confirmations of previous findings, the other 18 had never been connected to metal hyperaccumulation. Four of the genes with the highest transcript level in *A. halleri*, *HMA4*, *ZIP9*, *ZIP6*, and *ZIP3*, might exist as multiple genomic copies. The genes *ZIP9, ZIP6,* and *ZIP3*are members of the ZIP family of metal transporters and candidates for cytoplasmic metal influx in roots, while *HMA4* (At2g19110) has been characterized as a P_{1B}-type heavy metal ATPase in *A. thaliana* and is suggested to have a role in root-to-shoot Zn transport. The multiplication of these genes within the *A. halleri* genome could be an explanation for the high levels of expression, as has been recently shown for*HMA4*. All four of these genes are believed to have functions directly related to metal transport; thus, it would be fitting that they would be up-regulated in a hyperaccumulator such as *A. halleri*.

THE IMPACT

An efficient root uptake system, as well as root-to-shoot translocation of metals, is among the more important characteristics of a metal hyperaccumulator. Studies such as the one by Talke et al. (2006) have identified candidate genes up-regulated in the hyperaccumulator *A. halleri* relative to *A. thaliana,* including those involved in root-to-shoot translocation such as *HMA4. A. halleri,* a known Zn hyperaccumulator, also will hyperaccumulate Cd in aboveground biomass, most likely using components of the Zn (or iron [Fe]) transport. This double use of the transport system can be problematic in biofortification efforts of food crops in which the accumulation of toxic compounds such as Cd is detrimental. Thus, it is important to know and understand the details of the uptake process. The translocation of Cd from root to xylem was studied byUeno et al. (2008) and their findings confirm the use of components of the Zn transport system for Cd transport.

The root-to-shoot translocation was abolished by the metabolic inhibitor carbonyl cyanide 3-chlorophenylhydrazone, strongly suggesting an active transport system. HMA4, a P_{1B}-type heavy metal ATPase involved in Zn root-to-shoot transport, possibly plays a role in this process. Their study also found that Cd is translocated in the xylem predominantly as aqueous free ions rather than being complexed with citrate, which is believed to play an important role in the loading of aluminum and Fe.

The flow of the metals through the plant was the focus of a study by Waters and Grusak (2008), with the goal of identifying tissues that served as metal sinks for transport to seed to aid in the biofortification of crop plants by ultimately increasing seed metal content. Waters and Grusak (2008) performed whole-plant partitioning to determine the mineral content of tissues during growth stages in three wild-type *A. thaliana* ecotypes (Columbia-0, Landsberg *erecta,* and Cape Verde Islands) and one mutant (Columbia-07"*ysl1ysl3*) that has low copper, Fe, and Zn levels in seeds most likely due to lesions in genes proposed to be involved in metal transport from the vascular tissue. Although ecotype-specific differences were seen in metal partitioning at the tissue level, an increase in shoot metal content over the course of the plants' life cycle was common in all three. They were also able to determine that in *A. thaliana*, silique hulls are a mineral source for seeds. In the *ysl1ysl3* double mutant, there is an increased leaf-to-stem ratio of Cu and Zn, suggesting that metals are not partitioned properly in the plant. The authors propose that this might be due to a decrease in the mobilization of these metals from or translocation through the fruit hulls in the double mutant. Together, this suggests that, in the case of Cu and Zn, these metals need to go through a source tissue (leaves) before being translocated to the seed sinks. *YSL1* and *YSL3*, both of which Talke et al. (2006) found to be more highly expressed in *A. halleri*, need to be functional for efficient mobilization and are potential targets for the increase of minerals into seeds.

The identification of the proteins involved in metal hyperaccumulation, such as in the study by Talke et al. (2006), is an important step in fully understanding the process of tolerance of high metal content in plants. This information can, and has been, used in other studies on metal accumulation and to further research on biofortification of plants.

3

Plant Transport of Nutrients and Alkalinity

Plants with a high organic anion and associated cation content have a greater capacity to recycle soil cations and alkalinity to the surface in plant residues. Thus these species are useful in an ecosystem to achieve an excess of alkali in the upper part of the soil profile by increasing the rate of oxidation of organic anions near the surface. Where this process achieves a pH significantly above 5.5, the leaching of a net alkaline solution downward helps to neutralise acid addition deeper in the profile. Among the agricultural species clovers and medics are higher in organic anions (75 to 150 cmol/kg) than high quality grasses (50 to 70 cmol/kg). Cereals and low quality grasses that are relatively high in cellulose are low (< 60 cmol/kg) in organic anions and ca-tions (SLATTERY et al., 1991). Some dicotyledonous weeds such as cape weed and Paterson's curse are high in organic anions. Among the tree species eucalypts and acacias are often low in organic anions (< 100 and often < 60 cmol/kg leaf litter) while other species such as poplars, plane trees and white cedars are particu-larly high (150 to 250 cmol/kg leaf litter). Limited data on Casuarina spp. indicates they may be in the high end of the eucalypt range and other species have organic anion concentrations between 100 and 150 (NOBLE et al., 1996). This factor should be taken into account when trying to design more sustainable agricultural and forestry ecosystems.

The effect of the major processes affecting the pattern of acidity in a profile have been modelled (HELYAR, 2001). This model demonstrates that the equilibrium soil pH profile developed under a given pattern of acid and alkali production is very sensitive to some factors that are subject to management. Examples are the organic anion content of the plant grown, the depth of leaching of ammo-nium before nitrification and of nitrate before absorption by plants.

Furthermore, where the production system results in net acid addition to the soil profile, a deep, highly acid profile will develop over time unless sufficient lime is applied to the surface to neutralise the acid added. Even in natural ecosystems that achieve very low or zero net acid addition, (no nitrate leaching and no export of organic anions from the system), a duplex soil with an acid sub-surface layer can form where plant recycling of alkalinity from the subsoil is low. This results in net acid addition to the surface soil and net alkali addition to the subsoil. The consequences for soil profile development and for soil fertility are profound.

MANAGING AGRICULTURALLY INDUCED SOIL ACIDITY

To stop a highly acid layer developing in some section of the profile lime application must equal alkali removal as the first step to establishing a fertile pH profile without a strongly acid layer any-where in the root zone.

One exception to this suggestion is where the acid and alkali additions are such that a highly acid layer is created at the bottom of the root zone and the pH is above 5.5 elsewhere. In this situation the acid layer is used to mobilise nu-trients from parent material and accumulate them in the main root zone. The soils under Brigalow vegetation may be an example of such a soil in Australia.

The Surface Soil Should be Limed to Ph 5.5 or Above

This will ensure that the net movement of alkali to the deeper layers within the profile. A soil pH greater than 5.5 will also remove most of the problems associated with soil acidity. Care should be taken not to increase the pH of the surface 10 cm signi-ficantly above 6.0 because this may induce deficiency in other plant nutrients such as zinc, boron and manganese in well weathered soils.

Where the soil is acid to depth it may take many years for the alkaline effect of the lime to move below 20 cm, especially for highly buffered soils. Changes in pH in the subsoil layers of an acidic soil have been measured for 12 years in a trial to the east of Wagga Wagga (HELYAR et al., 1997). This has confirmed that if the pH is maintained around 5.5 in the top 10 cm that the pH at 15 to 20 will increase.

There is opportunity to correct acidity in sub-surface layers where the exchange capacity of the sub-surface soils is positive. Here gypsum, that is far more soluble than lime, can be used neutralise acidity at depth.

$$> \mathrm{Al-OH} + \mathrm{SO_4^{2-}} \quad > \mathrm{Al-SO_4^-} + \mathrm{OH^-}$$

$$> \mathrm{Fe-OH} + \mathrm{SO_4^{2-}} \quad > \mathrm{Fe-SO_4^-} + \mathrm{OH^-}$$

Reduce the Leaching of Nitrate Nitrogen

In soils with a pH above 4.7 nitrate nitrogen (NO3-N) is the main form of nitrogen that is available to the plant. Nitrate nitrogen is formed from ammonium nitrogen (NH4-N) by the process of nitri-fication. It is easily leached as it is weakly adsorbed by soil materials and is highly soluble. When NO3-N is leached away from the point of nitrification it leaves that point more acid. If the NO3-N is taken up further down the profile this can increase pH at the point of uptake.

When the NO3-N is leached below the root zone and into the sub soil aquifers it leaves the soil more acid. The factors that affect NO3-N leaching in the Wagga Wagga district roughly in order of importance, qualifies the effect of each factor and indicates how the effect can be reduced, or increased. The absolute effect of each factor will depend on rainfall and is usually periodic. In wetter climates the effect may occur every year but in drier climates it may occur in some years.

Removal of Produce

Grain, pasture and animal products are slightly alkaline and their removal from a paddock leaves the soil more acid. If very little produce is removed, such as in beef production, then the system remains almost balanced. If, however, a large quantity of produce is removed, particularly clover or alfalfa hay, products that are rela-tively high in organic anions, the soil becomes significantly more acid. If the produce is sold off-farm, regular liming is the only way to maintain pH.

The rates of lime required to neutralise the acidification caused by removal of produce are given in Table. The acidification caused by removing hay or silage is neutralised if it is fed to livestock in the paddock where it was made and the waste products are distributed evenly over the paddock. Neutralisation of the acid added to forest soils as a result of biomass accumulation and export in logs should be considered as a component of sustainable management of forests. Otherwise long term acidification of the forest soil will occur. The rate of lime requi-red can be calculated from the amount of organic anions removed in the logs plus the amount in the standing biomass that remains after logging.

Table: The Amount Of Lime Needed To Neutralise The Acidification Caused by Removal of Produce (Slattery Et Al., 1991).

Produce	Lime requirement (kg/t of produce)
Corn and wheat	9
Soybeans and lupins	20
Meat	17
Grass hay	25
Clover hay	40
Maize silage	40
Lucerne hay	70

Use of Fertilizers

The acidification that results from using fertilizers depends on the fertilizer type. Some nitrogenous fertilizers, for example urea, have no effect on pH if all the nitrogen is utilised by the plant. Acidity will only result if the nitrate nitrogen produced from these fertilizers is leached.

Table. Acidifying Effect of Nitrogenous Fertilizers and Legume-fixed Nitrogen. Lime Required to balance Acidification

(kg lime/kg N) Source of nitrogen	0 % nitrate leached	50 % nitrate leached	100 % nitrate leached
High acidification			
Sulfate of ammonia,			
Mono-ammonium phosphate (MAP)	3.7	5.4	7.1
Medium acidification Di-ammonium phosphate (DAP)	1.8	3.6	5.3
Low acidification			
Urea			
Ammonium nitrate			
Aqua ammonia			
Anhydrous ammonia			
Legume fixed N	0	1.8	3.6
Alkalisation			
sodium and calcium nitrate	-3.61	-1.8	0

1 Equivalent to applying 3.6 kg lime/ kg N.

Other nitrogenous fertilizers, for example sulfate of ammonia, have an acid reaction and their use acidifies the soil even if fully utilised by plants. Additional acidification results if nitrate nitrogen is formed and leached. Calcium nitrate, potassium nitrate and sodium nitrate neutralise soil acidity if utilised by plants and have no effect if all the nitrate is leached, but they are expensive and have limited application.

Superphosphate has no direct affect on soil pH, but its use stimulates growth of clovers and other legumes. In pasture systems where most of the increased plant growth is returned to the soil, increases of soil N status and a build-up of soil organic matter occur, both of which increase soil acidity. Elemental sulfur is acidifying and requires 3.125 kg of lime for each kg of sulfur to neutralise its effect. This effect can be avoided by using products that contain sulfur in the sulfate form such as gypsum, potassium sulfate and superphosphate. Acidification of the soil can be reduced by avoiding the use of highly acidifying fertilizers such as sulfate of ammonia and mono-ammonium phosphate (MAP). Nitrogen fertilizer (including urea) that is pre-sown should be drilled into narrow bands to slow nitrification and subsequent leaching. Surface application of nitrogenous ferti-liser for crops before sowing, even if harrowed, can result in nitrate leaching and consequent acidification. Nitrate nitrogen formed from post-emergent applications is more likely to be utilised by the crop and will cause less acidification.

DEGRADATION OF PRAIRIE SOIL RESOURCES

"Despite its widespread severity and global impact, soil degradation remains an emotional rhetoric rather than a precise and quantifiable scientific entity." Soil is degraded as a result of processes that reduce its productivity. Such processes usually arise from poor management of the resource. Soils are degraded primarily by erosion, loss of organic matter, salinization and acidification. The first two processes are interrelated. Acidification is not a major problem except in areas where cultivation has been extended into previously forested areas. Salinization is more prevalent.

SOIL EROSION

Soil erosion reduces soil productivity through losses of nutrients, water storage capacity and organic matter. The losses in terms of productive capacity are significant though presently obscured by the application of fertilizer. Annual top soil losses in terms of lost productivity on the prairies are estimated to be in the millions of dollars. The present value of the accumulated annual losses is measured in hundreds of millions of dollars, the actual losses being difficult to quantify. Erosion has taken place over the history of cultivation on the prairies. Both wind and water erosion result in the removal of the finer soil particles, thus leading to compaction of the soil and poor soil tilth. Water erosion is a significant problem on long sloping fields, particularly those summer fallowed or left free of cover. The valuable top soil removed is deposited in depressions, road ditches and waterways giving rise to additional costs. Wind erosion is a major problem when there is little cover by crops or residues as in the spring months particularly during dry seasons. Not only are valuable nutrients lost but the soil deposited elsewhere becomes a problem. Indeed, seedling crops may be destroyed.

Dumanski et al have estimated the annual costs of erosion in the prairie provinces to lie in the range of 155-177 million dollars in the case of water and between 213-271 million dollars in the case of wind. Sparrow has expressed annual soil losses in different terms. He estimated the annual soil loss on the prairies by wind to be about 160 million tonnes and that by water to be 117 million tonnes, these losses from erosion greatly exceeding the rate of soil formation. The annual loss in potential grain production arising from soil erosion during the period of cultivation he considered to be 171 million bushels or 4.6 million tonnes of wheat. He argued that 15 per cent of the lost production could not be recovered by additions of fertilizer and estimated the value of lost production at $129 million annually, lending credence to the title of his report*Soil at Risk: Canada's Eroding Future.*

The physical effects of erosion are painfully evident on the knolls and sloping segments of the prairie landscape. It has led to a gradual evolution of farming practices from an initial use of the plow, through use of blade type cultivation to use of minimum tillage. This progression has enabled increasing amounts of trash cover to be maintained over time, thereby reducing erosion. Moreover, in more recent years the width of fields has increased as a reflecting the size of the equipment used, which resulted from the economies of size in farm operations. In addition, intensive tillage operations to incorporate herbicides and fertilizers were adopted. These latter events work against the soil conservation otherwise achieved through maintenance of trash cover. Fortunately, a reduction in the cost of suitable chemicals has made the use of chemical fallow more economical, thereby enabling adoption of minimum tillage practices. This change in practice has been driven by the need to minimize production costs, the protection of the land from erosion being a complementary benefit. At the same time, summer fallowing has largely become confined to the brown soil zone as rotations have been extended in the dark brown zone and continuous cropping has become general in the black soil zone. Losses from soil erosion can therefore be expected to be less extensive in the future than in the past.

Decline in Organic Matter

The main chemical constituents of soil organic matter are carbon and nitrogen. The former is the energy source for most soil microbes while the latter is one of the most important nutrients for plants. Other necessary components of organic matter are phosphorous and sulfur. Under the ecosystem prevailing previous to cultivation, mineralization and immobilization processes were closely integrated. Plant growth (immobilization) and mineralization of soil organic matter occurred simultaneously. Therefore losses of nutrients were minimized allowing for the accretion of organic matter over time. Carbon transfer through the soil is basic to the functioning of the soil-plant system. Primary productivity determines the amount of carbon entering the system while the amount leaving is controlled by soil micro-organisms. In natural ecosystems, the amount of carbon transferred by means of both processes appears equal.

Upon cultivation of the soil for the production of annual crops, productivity and mineralization are uncoupled, particularly when the fields are bare of vegetation. Total inputs of carbon to the soil are altered due to differences both in plant characteristics and to allocation of carbon for above and below ground production.

The practice of summer fallowing has hastened the loss of soil organic matter. Frequent tillage during summer fallowing has increased the rate of decomposition of organic matter while at the same time disrupting soil aggregation and increasing its susceptibility to erosion. There has also been a loss of the more readily available nitrogen.

The magnitude of the losses of organic matter under cereal production has been established as being in the range of 20 to 50 per cent in the cultivated layer according to the soil classification. McGill estimated in 1981 that organic matter losses in the topmost layer of soil averaged 41 per cent in the brown soil zone, 44 per cent in the dark brown soil zone, 49 per cent in the black soil zone and 36 per cent in the grey soil zone. Research has established that by improved cultural practices, the decline in organic matter can be halted and indeed reversed. Nonetheless, additional nitrogen is now required to maintain yields on many prairie soils. Uncertainty remains, however, as to how soils will behave under such changed cultural practices as snow trapping, minimum tillage, green manuring and chemical follow. This uncertainty arises because soil changes occur slowly and the effects of agronomic practices take years to become evident.

SOIL SALINITY

Even before cultivation, large areas of the prairies contained saline soils. However, an expansion of these areas has occurred under cultivation. The rate of increase in the salinized area is related in part to the increase in ground water over time. Salinization occurs as water containing soluble salts moves upward through the soil horizon. Evapotranspiration concentrates the salts in the soil solution to reach levels in the root zone that are detrimental to crop growth.

Any technique which removes water from the subsurface such as drainage, use of deep rooted perennials, reducing summer fallow or continuous cropping diminishes the danger of salinization. Once salinization has occurred, the remedy is to leach the salts out of the root zone. The most viable way to accomplish this cleansing is to use all available precipitation to grow crops and thus reduce the accumulation of ground water. This indicates the desirability of reducing the acreage in summer fallow on the prairies. Such a reduction would also lead to a significant decline in the leaching of nitrogen into the subsoil.

Salinization of dryland agricultural soil constitutes a major problem in southern and central regions of Alberta and Saskatchewan and in a few scattered areas of Manitoba. Approximately 2.2 million hectares of improved land are considered to be subject to secondary salinity. The economic impact of this salinity is said to be in the range of $104 - $257 million annually.

SOIL ACIDIFICATION

As indicated previously, acidification is not a major problem in other than a limited part of the prairies. Acidification can be overcome by application of lime. However, heavy applications of nitrogen fertilizers over time have been found to result in soil acidification. Consequently, acidification may become a significant problem in parts of the prairie region under intensive cropping practices which require major inputs of nitrogen fertilizer.

Assessment

While degradation of the soil resource can be reduced or halted by changes in cultural practices, unless these are demonstrated to be economic they are unlikely to be adopted even though desirable from the standpoint of society. Producers are more likely to be interested in maximizing profits in the short-run rather than be concerned with conservation of natural resources and protection of the environment. Any sizable shift in cropping practices by producers will depend on their ability to survive difficult economic times.

SOIL AND LAND INFORMATION

DEWNR manages the State Land and Soil Information Framework, an important knowledge bank of detailed, comprehensive information on South Australia's soils and landscapes.

What is Soil and Land Information used for?

Landscape and soil features associated with underlying geology, landform and soil type (together with climate) are major factors influencing vegetation growth, water movement and the suitability of land for a variety of uses. A better understanding of land and soil helps us appreciate why different natural and production systems can be supported in different areas.

Good knowledge of soil and land improves our decision-making capacity, and is used for:

- Natural resources management – planning and policy development; guiding investment priorities; identifying the location, extent and severity of land management issues and areas prone to degradation; targeting on-ground works; supporting whole-of-landscape sustainable land use and management decisions for both production and nature conservation
- Land use planning – identifying limitations and opportunities for particular land uses and developments, from agricultural to urban, including the identification of 'prime agricultural land'
- Biodiversity management – supporting whole-of-landscape environmental management planning; identifying suitable areas for habitat restoration; identifying threats to the environment
- Sustainable soil and land management – identifying threats to our soil and land resources (eg erosion, acidification, salinity), as well as opportunities for improving management and condition
- Sustainable water management – identifying soil and land conditions, land management practices and land uses that impact on water resources, as well as developing whole-of-landscape solutions
- Landscape modelling – developing conceptual models that answer specific questions, such as: environmental risk assessment (eg land salinisation); identification of biodiversity assets for protection (eg wetlands); land use potential (eg crop potential modelling); land management potential (eg suitability of areas for specific soil amelioration); infrastructure risk (eg underground cabling); and scenario modelling (eg modelling the potential impacts of climate change on land use, land degradation or soil carbon)

- Monitoring – as benchmarking information for monitoring programmes; informing what to monitor and where; extrapolation of monitoring results across the landscape
- Research and education – identifying where in the landscape particular research outcomes apply; education of land managers, advisors, policy makers, planners, industry and the community about better soil and land management, and associated planning and policy development.

WATER, AIR, SOIL, AND NATURAL RESOURCES

Pollution of air, water, soil, and sediments is a significant problem in the world today, and the monitoring, reduction, and remediation of pollution is a critical part of the work of environmental and ecological engineers worldwide. Atmospheric pollution from industrial facilities, power generation, and transportation directly impacts human health and aesthetics. Chemical and biological contamination of water sources (both surface water and groundwater) leads to negative ecosystem effects and disease. Contaminated soil from waste disposal affects plant growth and land usability. But none of these problems exists by itself — most contaminants will move from one environmental reservoir to another. Engineers seek to understand the behaviour of contaminants and design technologies and practices to clean up contaminated sites.

Also, when possible, engineers design processes that remove or sequester contaminants before they are released into the environment. Research and education at Purdue in Environmental and Ecological Engineering covers a wide range of issues, including remediation of contaminated soils and sediments, industrial and solid waste treatment, water and wastewater treatment, air pollution measurement and control, urban and agricultural air and water quality management, understanding the environmental fate of pollutants, and sustainable engineering. Purdue has particular strengths in water treatment, both of drinking water (for example, the development of new, cost-effective, low-energy demand, and portable technologies for water disinfection that could be applied globally to provide our growing population with safe water) and wastewater (including the ecological engineering approaches of constructed wetlands to effectively treat wastes).

SOIL - AN ESSENTIAL NATURAL RESOURCE

Soil is one of the most important and essential natural resources. Farmers and gardeners know this fact and go to great efforts to conserve it. They understand that when the soil is destroyed, then gardens will not be successful and good crops cannot grow. They also know that it takes a considerable amount of time for soil to form. Here we will look at elements that must come together in order to form soil.

Soil is defined as a covering over most of the earth's land surface. It is made of particles of rock and minerals, living things and the remains of living things. It takes thousands of years for soil to form just a few inches and for some parts of the country it has been less than that.

Soil is formed in two general ways:

- When weather conditions cause rock to break down.
- When soil is carried from one place to another.

How is soil made?

Air

Moving air, such as the wind blows sand against rock to wear it down.

Water

The force of running water can wear away rock. Rain and snow also help to break down rock into smaller particles.

Plants

Plants help the soil as they grow and when they die. As dead plants decay, it adds an organic material to the soil, which makes it more fertile.

Animals

The body waste of many animals helps to keep the soil fertile, as well as the remains of dead animals as they decay. Now that you've had a lesson on the importance of soil as a natural resource and how it forms, you are now a more knowledgeable gardener. Use your knowledge to continually improve your gardening skills the next time you dig in the dirt to plant your flower or vegetable garden. Use your knowledge to develop an even greater appreciation for a natural resource that man cannot make, but has a responsibility to conserve and preserve for generations to come.

NUTRIENT DEFICIENCY AND TOXICITY SYMPTOMS IN TREE FRUIT

Growers often look at their orchards and suspect that something is not quite correct. Listed below are general guidelines for diagnosing symptoms that result from deficiencies or toxicities of certain elements. A tree may be deficient in these elements not only because they are scarce but also because an excess of other elements has prevented a balanced uptake of essential elements. The guidelines are not precise, and drastic changes in fertilizer practices should not be made based on a visual assessment. If you suspect a nutrient deficiency or toxicity, a leaf analysis should be conducted to confirm the visual symptoms. Information that follows was adapted from Childers, Shear and Faust, and Stiles and Reid. Elements required by plants are broken down into two broad categories: macronutrients and micronutrients. Plants need both to grow, flower, and fruit naturally. The distinction between the two groups is the quantities required by the plant. Macronutrients are needed in a larger amount and are usually expressed as per cent dry weight. The micronutrients are required in smaller quantities and are usually expressed in parts per million.

Macronutrients

Nitrogen

- *Deficiency:* Symptoms appear as reduced top growth with short spindly shoots that have pale green to yellow leaves. Generally speaking, symptoms first become evident in the older leaves at the base of shoots. In stone fruits deficient leaves appear reddish and may exhibit a "shothole" effect as the condition worsens. Fruits, especially stone fruits, tend to be smaller and to mature earlier.
- *Toxicity:* Symptoms appear as an excessive amount of shoot growth accompanied by dark green foliage and delayed leaf drop in the fall. As nitrogen increases above the optimum, fruit colour is reduced and maturity is delayed. Red cultivars are less red, and yellow cultivars tend to remain green. In apples and pears flavour and storage life are reduced. Besides direct toxic effects, other physiological problems, such as corking and bitter pit, can occur in apples and pears because of higher nitrogen levels that are not in balance with calcium levels.

Magnesium

- *Deficiency:* In their severest stages, symptoms may look similar to the marginal scorching associated with potassium deficiency, although this is very rarely observed in Pennsylvania. More characteristic is a fading of the green colour at the terminals of older leaves, progressing interveinally (between veins) towards the base and midrib of the leaf and giving the typical "herringbone" appearance. In pears, dark purplish islands of tissue surrounded by chlorotic bands may develop in the interveinal areas. As the growing season progresses, symptoms develop on progressively younger leaves and the older leaves fall off.
- *Toxicity:* Symptoms of excessive magnesium levels are not specific but usually appear as a deficiency of either potassium or calcium.

Micronutrients

The following elements are classified as micronutrients because they are required by plants in smaller quantities. With the exception of manganese, toxicities are very difficult to diagnose visually. Deficiency symptoms tend to be characteristic of the lack of a particular element.

Iron

- *Deficiency:* Iron deficiencies are very common in plants. Initial symptoms are a loss of green colour in the very young leaves. While the interveinal tissue becomes pale green, yellow, or even white, the veins remain dark green. New leaves may unfold completely devoid of colour, but the veins usually turn green later.
- *Toxicity:* Although rare in the field, an excess of iron usually produces symptoms similar to those of manganese deficiency.

Manganese

- *Deficiency:* Symptoms begin as chlorosis between the main veins starting near the margin of the leaf and extending towards the midrib. Symptoms can often be confused with those of iron and magnesium deficiencies. But unlike magnesium deficiency, manganese deficiency symptoms seldom develop so far as to produce interveinal chlorosis, the chlorosis normally being confined to leaf margins. The other distinguishing characteristic is that manganese deficiencies appear on the youngest leaves first, and the finest leaf veins do not remain green as they do with iron deficiencies.
- *Toxicity:* "Measles" is a disorder of apples, especially Delicious and Jonathan. It is caused in part by an excess of manganese accompanied by low calcium levels.

Boron

- *Deficiency:* In most fruit crops, boron deficiencies show up in the fruit before appearing in the leaves. Symptoms in apples and pears are similar: gnarled, misshapen fruit caused by depressions usually underlaid by hard corky tissue. This symptom is often confused with bitter pit or corking caused by calcium deficiency. Researchers disagree whether the two symptoms can be told apart visually. Boron deficiency might be distinguished from bitter pit by the presence of pitting from the peel to the core, whereas in bitter pit the pitting usually occurs only at the calyx end and only very close to the skin. In some instances of boron deficiency, the entire surface is covered with cracks that have callused over, producing a russeted appearance. In plums, the symptoms appear as brown sunken areas in the fruit flesh, ranging in size from small spots to almost the whole fruit. The fruit usually colours earlier than normal, and falls. Gum pockets may also form in the flesh. In peaches, the flesh adjacent to the pit develops brown, dry, corky areas, and some fruits may crack along the suture. The most typical vegetative symptom is the death of the terminal growing points, resulting in a "witch's broom" appearance.

- *Toxicity:* Symptoms in apples include dieback of twigs, greatly enlarged nodes on 1–and 2–year–old twigs, early fruit maturity, internal breakdown, and dropping of fruit. Foliar symptoms occur first on the older leaves and include a yellowing along the midrib and the large lateral veins. In peaches vegetative symptoms include necrotic lesions on leaves, crinkling of margins and tips of leaves, reduced flower bud formation and set, and pit splitting.

Copper

- *Deficiency:* Younger leaves appear stunted or misshapen, narrow, and slightly elongated with wavy margins. There may be some terminal dieback. Copper and zinc deficiencies often occur together and are aggravated in soils that have a high pH.
- *Toxicity:* Symptoms are almost nonexistent under orchard conditions, but when present they may resemble those of zinc deficiency.

Zinc

- *Deficiency:* Symptoms have often been described as a "rosetting" of leaves or "little leaf." Newly developing leaves are smaller than normal. Reduced shoot elongation keeps them close together, resulting in the rosette appearance. In severe cases, older leaves may drop, resulting in a more pronounced rosetting. In the early spring, the observer might notice a delayed foliation of lateral leaves on last year's shoots. This symptom has been confused with that of winter injury, but the distinguishing characteristic is that winter injury will also produce browning of the cambium.
- *Toxicity*: Symptoms are rare and most likely are masked by secondary symptoms resembling those of other micronutrient toxicities.

FOLIAR APPLICATION OF NUTRIENTS

The most efficient way to apply nitrogen, phosphorus, potassium, and magnesium is by ground application. Foliar applications of these elements should be viewed as temporary or emergency solutions only. Boron, zinc, copper, and manganese can be added by either foliar or ground application. The foliar method is usually preferred because very small amounts are applied per acre. Since formulations differ by manufacturer, it is essential to check compatibility of the material when mixing. Foliar application of calcium to control cork spot and bitter pit is dealt with in Cork Spot and Bitter Pit Fruit Disorders.

FERTILIZER APPLICATION

Nitrogen

Nitrogen requirements are best determined by tree growth and performance. Soil tests for nitrogen have not proven useful in determining tree needs. As mentioned previously, nitrogen levels in a foliar analyses should be higher for young nonbearing trees. Leaf nitrogen levels also tend to be higher in samples from trees carrying heavy crops. Biennial bearing trees in their off year or trees with a light crop generally have lower nitrogen levels. This is probably related to the inverse relationship between shoot growth and fruiting. Nitrogen cycle and its availability is a complex situation. Nearly 80 per cent of our atmosphere is composed of nitrogen. Some can be precipitated out during thunderstorms and rain events.

Additionally, much nitrogen is recycled when organic matter decomposes. It has been estimated that for every 1 per cent of organic matter in the soil as much as 20 to 45 pounds of nitrogen per acre per year can be made available for plants to absorb. In addition, much of the nitrogen in the leaves can be remobilized back into the tree to be stored in the structural portions of the tree. This stored nitrogen is what supplies the initial tree growth in the spring. The larger the tree, the more potential nitrogen the tree can store. Conversely, this means that the new smaller trees on a per–tree basis will store less nitrogen. In apples, we have a better understanding of nitrogen levels based on cultivar. Cornell University researchers suggest the following nitrogen levels in mature trees based on cultivar: 1.8 to 2.2 per cent nitrogen: Soft cultivars such as Cortland, Gala, Golden Delicious, Jerseymac, Jonagold, Jonamac, Jonathan, Macoun, McIntosh, Mutsu, Paulared, Spartan, Tydeman's Red, and other early ripening cultivars. 2.2 to 2.4 per cent nitrogen: Delicious, Empire, Idared, Liberty, Melrose, Rhode Island Greening, Rome Beauty, Stayman, York Imperial, and other varieties.

Application Rates

As a general rule, young trees may require about 0.01 to 0.04 pound of actual nitrogen per year of age up to 0.3 pound actual nitrogen per tree at maturity. These general levels should be adjusted up or down, depending on pruning, size of crop, cultural practices, indices presented above, and leaf analyses. Young nonbearing trees should grow up to 24 inches annually. Mature and heavily bearing trees should grow less, with apple and pear trees growing 12 to 18 inches; apricot, cherry, plum, and prune trees 15 to 18 inches; and peach and nectarine trees 18 to 24 inches.

Form of Nitrogen

Fruit trees respond to any form of nitrogen fertilizer. Avoid using ammonium sulfate, urea, or ammonium nitrate if pH is below 6.0. Urea sprays may help increase fruit set but will not supply trees with all the nitrogen they require. Pear trees do not respond well to urea sprays. Applications of urea sprays to stone fruits are ineffective during the growing season. Calcium nitrate sprays should be avoided to discourage corking.

Time of Application

Traditionally, we have said that nitrogen should be applied no later than 4 to 6 weeks before bloom. Recent research in the Pacific Northwest, however, indicates that application at this time results in the nitrogen chiefly accumulating in the foliage of the trees with little going into the flowers and developing fruitlets to help fruit set. Studies have shown that the majority of the nitrogen used to set fruit comes from the reserves within the tree and spring–applied nitrogen does not reach the developing fruit in time to be effective.

In studies in apples in Oregon, late–summer to early fall nitrogen application was found to be preferentially translocated to the roots and the flower buds where it was mobilized from the roots and available in the flower buds next spring to increase flower strength and fruit set capabilities. The theory that nitrogen should not be applied after the middle of July may be incorrect; however, we have no sound research on this different application timing in Pennsylvania. Until we have adequate information, the safest method is to continue your past fertilization practices or to experiment only on a small scale.

Phosphorus

Phosphorus is important to plant growth because it is a catalyzing agent that induces metabolic reactions. It permits the plant to use nitrogen and to develop seeds. Failure of seeds to set often results in abortion of the young fruit and in misshapen fruit. Preplant soil test recommendations are made to attempt to raise soil levels to 100 pounds available phosphorus per acre.

Application Rates

Rates should be based on either leaf analysis or soil analysis. Soil levels of less than 100 pounds phosphorus per acre indicate that an application of phosphorus is needed. Leaf analysis levels of less than 0.18 per cent in apples and pears; 0.15 per cent in peaches and nectarines; 0.23 per cent in cherries; and 0.09 per cent in plums indicate a need for phosphorus. Form of phosphorus: The effectiveness of special forms of phosphorus under Pennsylvania conditions has not been documented.

Time of Application

Application may be made anytime during the year in established orchards. Preferably, applications should be broadcast before the trees are planted and turned under with the previous crop. This aids in getting phosphorus down into the root zone.

Potassium

Potassium is believed important for maintaining water turgor in leaves and for functioning in the opening and closing of stomates. It is available in the soil as a cation; at excessive levels it competes with calcium and magnesium for uptake by the plant.

Application Rates

If the potassium level in the foliage is over 1.5 per cent, a response to potassium application is doubtful regardless of the soil test, unless leaf nitrogen or soil magnesium is too high. When soil test values exceed 4.5 per cent base saturation of potassium, then no additional potassium is needed. Values lower than this call for an application. Form of potassium: Any added value of one type of potassium fertilizer over another has not been determined. Time of application: Same timing as for phosphorus.

Calcium

Calcium plays a vital role in reducing the incidence of corking and bitter pit. Low soil–test values are often found in very sandy or shaley soils, but such values usually do not indicate a need to apply calcium or lime. Calcium in the form of limestone is important for maintaining soil pH.

Improper soil pH can lead to deficiencies or toxicities of other nutrients. The need for lime is best determined by a soil test.

Calcium, like potassium and magnesium, is available to the plant as a cation, and an excess of potassium or magnesium can reduce calcium uptake. Application rates: Soil application depends on soil pH, buffer pH, and depth of pH change desired.

Calcium chloride applied as a foliar spray is recommended to prevent corking and bitter pit in the fruit. Form of calcium: Growers should base their limestonepurchase decisions on the price per ton of calcium carbonate equivalent, including spreading costs. Most limestone contains some impurities, the most common being magnesium. Continual use of dolomitic or high–magnesium lime can cause problems in a tree's uptake of calcium.

Dolomitic lime should not be used unless the soil test or the leaf analysis indicates a need for magnesium. Time of application: In preplanting situations lime is most effective when broadcast and incorporated at least 6 to 12 months prior to planting. In established orchards any time is suitable, although postharvest or early spring applications allow rains to move lime into the soil. Regardless of how much rain accumulates, the process of raising the soil pH is slow, with the effects of lime moving downward at about 1 inch per year.

Magnesium

Leaf analysis values below 0.2 per cent in apples and plums, 0.3 per cent in peaches and pears, and 0.49 per cent in cherries suggest that trees may respond to applications of magnesium. In soil tests, when the ratio of the percentage base saturation of magnesium to potassium is less than 2.0, magnesium applications may be needed.

Application Rates

Foliar applications of magnesium sulfate at 10 pounds per acre should be viewed as quick but temporary measures. A more permanent solution is to apply a magnesium containing limestone at rates recommended in the soil analysis. Form of magnesium: Magnesium is normally found in all but the purest limestone. The Penn State soil test results give a separate recommendation for magnesium. Growers should examine the purity of the limestone to determine whether the percentage of magnesium contained in it will also satisfy the magnesium requirement.

Time of Application

Magnesium sprays are effective only when applied during the growing season. Applications may be single or split (*i.e.*, applied at petal fall or in the first two cover sprays). Foliar sprays appear to be most effective when applied separately from pesticide applications. Ground applications can be made when lime is applied.

Boron

Boron is a nutrient often found to be low or deficient in orchards. Deficiency symptoms are observed more often in apple, plum, and pear trees than in peach and cherry trees. In apple trees, a deficiency may be expressed as internal cork in the fruit, as a dieback of shoots, and as bark necrosis. In pear trees, a deficiency may be expressed as internal cork, withering of blossoms, or poor fruit set. In peach trees, toxicity symptoms appear as necrotic lesions on leaves, crinkling of margins and tips of leaves, reduced flower bud formation and set, and pit splitting. Tissue test values less than 35 ppm in apples indicate a shortage of boron. Values between 35 and 60 ppm indicate sufficiency. In the soil, values less than 0.5 ppm are low, while values between 0.5 and 1.0 ppm indicate a sufficient level. Caution: Excessive boron can be extremely toxic. Leaf values of over 80 ppm or soil values of over 1.0 ppm are excessive.

Application Rates

In orchards where boron is low, apply 0.8 to 1.6 pounds per acre of actual boron (4–8 lb/A of Solubor 20.5% B) in two separate sprays at bloom, petal fall, or first cover; or apply a single postharvest foliar spray of 1.6 pounds per acre of actual boron. For trees more than 3 years old, apply boron annually to the soil. Apply 0.12 pound of actual boron per acre for 4–year–old trees. For each additional year of tree age up to 16 years, increase the rate by 0.02 pound per acre of actual boron. Use the lower rate as a maintenance programme when no leaf analysis has been made. Eight pounds of Solubor per acre is recommended for proven cases of low boron. Form of boron: Boron is most effective when applied as a foliar spray. Special formulations such as Boro–spray or Solubor should be used. Agricultural borax is adequate for ground applications. Time of application: Soil applications can be made anytime. Foliar applications can be made during bloom or postharvest while leaves are still green and active on the tree.

Copper

Copper is a catalyzing element in plant metabolic reactions. Foliar analysis is best for determining copper deficiency. Values below 5 ppm indicate the need to apply copper. Application rates: When leaf analysis indicates a deficiency, 4 to 6 pounds per acre of copper sulfate is recommended. Form of copper: Copper sulfate (22% Cu) is the most readily available and cheapest material to use. Time of application: Foliar applications should be made during the dormant season or after the fruit is off and while the leaves are still active and green. Soil applications can be made anytime.

Zinc

Deficiencies of zinc are becoming more common in Pennsylvania. Small leaves clustered on the end of shoots indicate zinc deficiency but may be confused with winter injury. Zinc applications are recommended when leaf levels are below 20 ppm. Foliar applications of zinc sulfate should only be made during dormancy or postharvest. Do not apply zinc sulfate with oil or apply within 30 days of the application of oil.

4

Transpiration to Land Plant

INTRODUCTION

Transpiration exists in all land plants. Whether it be an advantage or a disadvantage, the fact remains that it is quite unavoidable. The absorption of carbon dioxide from the air by the leaf is a prime necessity for the synthesis of carbohydrates by the living plant, and it follows that there must be some suitable mechanism of gaseous exchange in the organisation of the plant body, and this exchange is principally performed by the stomata of the leaf. As a corollary, it follows that if stomata are present, water losses must occur.

The aerial portion of the plant soma is covered with cuticle, and were this cuticle to extend over the whole surface of the plant, it would be unable to carry out its normal function of carbon dioxide absorption and, of necessity, its metabolic activities would cease.

In the case of ordinary mesophytes the temperature of the leaves is not much higher than that of the surrounding atmosphere, usually about a degree or so. In high winds the temperature may even fall below that of the air. Even when leaves are cut the temperature only rises one or two degrees from "wound-shock."

It must also be remembered that in the case of the ordinary type of dorsiventral leaf with its flat surface, there is a very large area exposed to the air which would tend to prevent any rise in temperature should external conditions of drought and heat arise. The most exact work on this subject has been carried out by Clum using leaves of *Fuchsia speciosa, Brassica oleracea* and *Syringa vulgaris.*

The temperature of the leaves was determined by means of thermo-couples under a number of experimental conditions. The leaves were always found to be at a higher temperature than that of the surrounding air.

The effect of shading was found to cause a sudden drop in the temperature; further the direct effect of sunlight was very much marked, and the particular angle at which the rays struck the leaf had an effect on its temperature. In no case, however, could he find any correlation between leaf temperature and transpiration.

A second claim that has been put forward to show thatjrans-piration is an advantage to the plant is that it rids it of excess water and allows of the rapid absorption of salts. A considerable amount of evidence was at one time forthcoming in support of this view, but it has since been shown to be erroneous.

The process of transpiration has nothing to do with the absorption of salts; the two processes work independently of one another. The absorption of salts is a question of the salt and water equilibrium of the cell, and is not connected with the evaporation of water from the leaves. Moreover, the concentration of salts in the plant is very much higher than the concentration of the corresponding salts in the soil.

An early experiment of Ilassel-bring bears on this problem. He grew tobacco seedlings in the open where transpiration would be normal, and a second series under calico to reduce transpiration. At the end of a given time the ash-content of the two sets was determined, and was found to be identical in both series.

The shade plants had absorbed 35 litres of water and the sun plants 46 litres or 30 per cent. more. The amount of mineral matter absorbed by the plants was therefore not proportional to the transpiration, nor did the plants with the lower transpiration suffer with regard to salt absorption. Later experiments by other investigators have given substantially the same results.

Muenscher grew barley in water culture, and, by suitable arrangements, was able to cut down the transpiration rate. He ascertained that there was no difference in the ash-content of the series grown under conditions which reduced transpiration; the ash-content could only be correlated with the growth of the plants. Mendiola found that in tobacco a reduction in the amount of transpiration led to an increase in the dry-weight and a decrease in the ash-content. On the other hand, his results support the view that there is no proportionality between the amount of transpiration and the amount of salt absorbed by the plant. It is quite possible, however, that transpiration might increase the rate of transport of salts from one part of the plant to another, once the salts had effected an entrance.

The rate of the transpiration stream is rapid, while the passage of salts from cell to cell is slow. It has been shown that salts are present in the xylem, and it may well be that the transpiration stream conveys salts from the lower to the upper regions of the plant. There may be a critical rate of the transpiration stream below which the water fails to carry the salts in the xylem with any rapidity.

THE EFFECT OF LIGHT ON TRANSPIRATION

Although a great deal has been done in the way of discovering the effect of stomata on the rate of transpiration and also a number of facts and figures arc available for the transpiration of the leaf as a whole, very few attempts have been made to ascertain the direct effect of changing external conditions on the mesophyll alone, although it is from these cells that the main water loss occurs.

The action of the mesophyll lias in most cases never been separated from that of the stomata. The difficulty is to remove the action of the stomata. It was suggested by Knight that leaves from which the epidermis had been stripped might be employed, but the method is drastic, and there is the further danger of the drying of the mesophyll cells except in atmospheres of high humidity. The other method was that employed by F. Darwin.

The essentials of this method are to exclude the action of the stomata by stopping them up with some substance like vaseline and then placing the intercellular spaces in contact with the atmosphere by means of slits in the leaf tissue. Such slits were usually made with a scalpel between the principal veins of the leaves so as to interfere as little as possible with the water supply.

Using this method, Darwin attempted to determine the effects of changes of light and humidity of the atmosphere on the mesophyll tissues of the leaf.

He came to the conclusion that the leaf increased its transpiration by amounts that varied between 10 to 100 per cent, of that in the dark. An increase of the latter order is certainly most remarkably high. To bring about such a doubling of the evaporation rate from a moist surface in air of constant evaporating power would require an increase of temperature of at least 10°C.

From a careful consideration of Darwin's data, Henderson was able to ascertain two possible sources of error. Firstly, Darwin's observations were usually begun immediately after smearing the leaves with vaseline and slitting, thus making no allowance for "wound-shock response," and secondly, external conditions were seemingly quite uncontrolled and largely unknown. The plants were in some cases removed from the window of the laboratory into the dark-room, apparently without accurate observations being made of humidity and temperature. Henderson, by means of a highly ingenious piece of apparatus, has repeated and extended Darwin's observations. The plants used in these experiments were *Hedera helix*, *Eupatorium adenophorum* and *Aster*. It was found that the transpiration at first rose to a high value after slitting, and then settled down to a steady rate in the course of an hour or so. The cause of this strange behaviour was not ascertained. It was during this preliminary unsettled period that all Darwin's readings were taken.

Since all his experiments were begun in the light and the plants removed to the dark-room when the maximum was reached, it was not surprising that a drop was shown in the transpiration rate in the dark, and it is also of significance in this connection that when the plants were again placed in the light, the second curve nowhere reached the same value as the first. Had Darwin started his experiments in the dark, then in all probability his results wouldhave been reversed.

It was shown by Henderson at the conclusion of his experimental work that light does have a small direct effect upontranspiration from the mesophyll; the order of increase being from 4 to 5 per cent, of that in the dark. The exact action of the light on the mesophyll is unknown. Henderson advanced the view that it may be due to the presence of unknown biological complexes, and is possibly in the nature of a photo-chemical reaction.

It cannot, however, be said that the explanation is of a very helpful nature. In any case this action of the light shows that transpiration is not solely a simple process of evaporation. It is possible that a secretion of water from the mesophyll cells of the leaf takes place; the cells actively secreting water or some solution on their outer surfaces. Here again, it must be confessed that the word "secretion "in reality only acts as a cloak for our ignorance of the chemical and physical phenomena underlying the whole process.

THE TRANSPIRATION CURRENT

The question of how water is conveyed from the roots to the leaves of plants has long exercised the minds of plant physiologists, and a number of theories have been advanced to account for the process. In some plants, such as the Blue Gums of Australia, which are often as much as 300 feet high, the question of water transport becomes a serious one. It has been conclusively shown that the path of the current is through the xylem. It is easy to demonstrate this fact by simple experiments. If a cut shoot be placed in a weak solution of such a dye as eosin, the dye will only be found to be present in the trachea'.

Again, a ringed tree will continue to transpire for a considerable time, while water plants which transpire but little, have only weakly developed woody tissues. There is need here of a dynamic and not a static explanation of the process; in other words, an explanation that will adequately account for the continuous drive of water up the cavities of the xylem.

Such phenomena as evaporation, capillarity, atmospheric pressure and root pressure have all been invoked to account for the ascent of sap. Capillarity and atmospheric pressure are both, static processes; after water has risen a certain distance, no further rise can occur.

Root pressure is a dynamic process, but it, however, is low in summer, when transpiration is at its maximum, so that root pressure will not adequately account for the ascent of sap in plants. It is, therefore, not surprising that the alder investigators fell back on vital theories to account for the rise of the water current, and assumed that some kind of pumping action was initiated by the living cells in the xylem. such as the xlyem parenchyma.

Strasburger's classical experiments conducted many years ago, showed quite conclusively that water would still ascend in plant stems when the living cells had been killed, and thereby sounded the death knell to any vital theories, and physiologists were forced to fall back on purely physical explanations.

TRANSPIRATION STREAM

Theory originally propounded by Dixon and Joly in the 'nineties, that the ascent of sap was due to the cohesion of the water columns in the cavities of the xylem and was drawn off from above by the leaf, has been very fully dealt with by Dixoifln his monographs—Transpiration and the Ascent of Sap, and the Transpiration Stream. The theory Dixon has put forward and supported by a considerable amount of quantitative and qualitative data, has rather tended to overshadow all other work on the matter.

The chief disadvantage of the theory is that it explains too much. Briefly stated, the Dixon and Joly cohesion accounts for the ascent of sap by taking into consideration thte fact that water possesses strong cohesive properties and that quite considerable pressures may be developed in this way, in fact, pressures as high as 300 atmospheres have been recorded.

According to Dixon, then, the ascent of sap takes place as follows: the water in the conducting tracheae hangs in fine columns cohering to the walls of the tracheae. From this initial condition we can proceed to a consideration of the passage of water from the soil through root and stem to the leaves. Through the suction pressure of the root-hairs and cortical tissue, water is removed from the soil and passes into the xylem.

By means of its cohesive properties the water adheres to the wall of the cavities of the xylem in fine columns, and from these columns, which are in a high state of tensile strain, it is withdrawn by the mesophyll tissues of the leaf. These hanging columns of water can exert a pull equal to that of a fine steel wire, but they can only do this while they are attached to the walls of the containing tracheae.

The initial force in the process is exerted by the root. Although the cohesion theory of the ascent of sap explains in a remarkable manner a large number of facts that have been observed about the transpiration stream, in actual practice it is difficult to devise crucial experiments which would definitely settle the truth or otherwise of the theory.

From the Dixon point of view the matter is reversible, that is to say, the plant should react equally well if it were placed upside down. It must be remembered however, that it would be almost impossible to obtain the same contact between the particles of the soil and the leaves as exists between the root-hairs and the soil. The main criticism that has been levelled at the theory is the presence of air-bubbles in t,he wood.

The presence of air-bubbles immediately destroys the stability of the hanging columns and would cause their collapse. It is impossible to state definitely whether the air-bubbles seen in cut sections are not brought there by mechanical manipulation, or whether their presence is natural in the wood of the living plant.

The investigations of Bode seem to indicate that there are no bubbles in the xylcm normally, and that even in the last stages of wilting the water-content of the tracheae is continuous. Dixon accepted the presence of bubbles in the xylem. The difficulty is to remove them. The suggestion has been made that perhaps root pressure in the spring is sufficiently strong to get rid of them.

But when the whole plant is transpiring rapidly, a negative pressure is set up in the wood, and the bubbles are then under a negative pressure; when a positive pressure is introduced they will suffer compression. A further difficulty that arises from the Dixon point of view is the fact that tracheids should be more efficient than vessels for the conductipn of water.

In Dixon's own words: The salient feature of the wood is the subdivision of the waterways by an immense amount of transverse and longitudinal divisions into minute compartments the vessels and tracheids. For the system the function of which is the conducting of water this is evidently the most unexpected configuration.

It is true that the partitions are permeable to water; but when a considerable distance has to be traversed the sum of the distances opposed by the walls becomes very important. It is evident from the persistence of walls in the development of the water-conduits of a plant, introducing as they do an immense resistance to flow, is inexplicable on any view that regards the water being forced through the stem by some unknown force. The cohesion theory, on the other hand, gives a ready explanation, for it confers stability on the tensilely stressed transpiration/stream.

From this statement it is evident that tracheids with their numerous short compartments would give greater stability to a tensilely stressed stream of water than vessels in which the cross-partitions are relatively far apart. Yet, such being the case, it is surely a curious anomaly that our present dominant flora, the Angiosperms, have practically without exception adopted vessels in place of tracheids. The various objections raised in this discussion to the unqualified acceptance of the colics ion theory, as Professor V. H. Blackmail once remarked: "Somewhat dilute the milk of the cohesion theory."

It must be admitted that at the present time there is no adequate explanation of the ascent of sap, and although the cohesion of water possibly plays a considerable part in the matter, it is probably only a portion of the whole truth. In spite of the work of Strasburger and others, the vital theory of the ascent of sap has apparently still a hold on some minds. At the present time the most ardent exponent of such a theory is the Indian investigator Sir Jagardis Chuiulra Bose. According to this worker, water is driven through the plant by some kind of pulsatory activity of the living cells, these cells beyig identified with the innermost cells of the cortex.

The methods of reasoning which lead him to make these assertions arc remarkable and peculiar. Bose reviews the various theories that have been put forward from time to time to account for the rise of sap in plants and any view that the rise depends on purely physical grounds is considered to be quite untenable. The work of Dixon and Joly is summarily dismissed in three or four lines, Strasburger is held to have misinterpreted his results, and Bose then quotes a statement from Pfeffer's text-book of plant physiology.

"How and by what means the water is so rapidly transmitted even to the summit of the tallest trees has not yet been satisfactorily explained. It has unfortunately not yet been determined whether the aid of living cells is quite necessary." Apparently all that has since been discovered is of no consequence to Bose, who sets forth his views backed by a series of experiments chiefly interesting for the highly complex apparatus that was employed.

According to Bose, a pulsatory activity is initiated in the roots, affecting the absorption of water from the soil. A similar activity pumps water up through the stem from cell to cell and also injects water into the vascular tissue of the xylem along which it is physically transferred.

The water conveyed by physiological conduction and physical convection reaches the leaf, where it is distributed along the veins and their numerous ramifications, and is eventually excreted and transpired into the intercellular spaces, whence it finds its way to the atmosphere outside by evaporation. To establish this standpoint, Bose has taken the following experimental line.

Similarly, by means of his "crescograph," Bose asserted that growing organs also exhibit an autonomous activity of a pulsatory nature. Having convinced himself of the truth of these two statements, Bose further stated that the ascent of sap is modified in the same way by different conditions such as anaesthetics. Such being the case, he had no diiliculty in formulating his theory that the ascent of sap is effected by the rhythmic activity of living cells.

It should be noted," as Dixon has pointed out in his admirable criticism of this work *The Transpiration Stream*, "the peculiar verbal transitions which often form an essential part of his arguments. Here alternate contractions, varying from one to four minutes or so, become ' pulsations,' and these glide easily into ' rhythmic activity.' An experiment by Bose in support of his views is quoted here.

He employed an electrometric method depending on his idea that "the electric condition of the tissue undergoes a definite variation under changes of turgor; a diminution of turgor introduces an electric change to galvanometric negativity, while an increase in turgor, on the other hand, an electric change to galvanometric positivity." Now a plant under drought has its turgor diminished throughout its length. If then, two electrical contacts are made, A with the stem and B with a distant leaf, the electric condition of the two points will be more or less similar.

Addition of water to the soil, however, will cause an ascent of sap, which will reach the lower contact first and will induce an increase of turgor at this point. This will at once be signalled by a sudden change to positivity at A. This experiment was repeated by Dixon, aided by Dr. J. H. Poole, using a galvanometer which gave a deflection of 1 mm. for 0-004 volt. The plants employed were *Sparmannia africana* and *Pelargonium zonale*. Although the soil round each was dry, neither showed any signs of wilting. "

After connection had been made as described by Bose, a slow movement of the spot of light was recorded. A few minutes later the spot came to rest at 19-7 mm. to the left of zero. This, as the instrument was arranged, indicated that the point A was positive relative to B. When observation showed that the spot was sensibly stationary, the earth in which the plant was growing was irrigated with water at about 25°C. Immediately the spot moved to 17-0 mm. and slowly crept back to 19-3 mm., which it reached two minutes later. During the next thirty minutes it crept slowly to 21-0 mm. During all this time nothing like the to and fro motion, corresponding to Bose's pulsations, was observed, nor was there any sudden deflection.

Point A was 3 cm. above the surface of the soil. The temperature of the room was 14-0°C." Similar results were obtained for *Pelargonium zonale.* Dixon also failed to confirm Bose's statement of a rhythmic change of potential in the cells of the stem by means of Bose's method of the electric probe. This "electric probe "consisted of a platinum wire coated with celluloid and enclosed in a glass tube.

This glass tube was fixed firmly in the body of a microscope, so that the sharp point of the wire, from which the celluloid had been stripped, projected from the glass tube and occupied the position of the objective. With this instrument Dixon attempted to repeat Bose's results. The body of the microscope was placed horizontally and the vertical stem of the experimental plant was firmly clamped to the stage.

In this way the point of the probe was gradually advanced through the tissues of tin stern by the action of the micrometer screw of the microscope. By means of mercury in the heathing glass tube, connection was made between the probe and one pole of a galvano-meter, while the other pole was connected with another platinum contact embedded in the tissue of a morbid leaf.

The circuit was completed through the plant. With this apparatus and using a highly sensitive galvanometer giving a deflection of 1 mm. for a current of 10" amperes, Bose recorded that the galvanometer showed alternate deflections in opposite directions when the probe had advanced a certain distance inwards.

The period of this alternation varied according to circumstances from 13-5 seconds to about three minutes. The amplitude of the vibrations varied between 10 mm. and 66 mm. Dixon, with a galvanometer giving a deflection of 1 mm. for a current of 5 X 10 amperes, and using *Pelargonium zonale, Impatiens sultani* and *Chrysanthemum sinense,* found that there was a slight "kick "every time the probe was advanced 0-1 mm. into the tissue at intervals of ten minutes, but in no case was there any regular vibration recorded.

Bose considered that the osmotic activities of the leaf cells are too slow to account for the rapid rise of water in plant stems. "He quite fails to realise that slow osmotic activities spread over the large area of the bundle terminations would cause a comparatively rapid rise of water "(Dixon).

Bose's criticisms of Strasburger's work are entirely beside the point. He stated that the living cells above the dead portions of the stem are able to pump water through the vessels, and thus "unconsciously admits the substance of the cohesion theory ". Benedict has determined the applicability of Hose's theory by comparing the maximum rate of flow which is possible according to his theory with the known rate of flow.

The cross-sectional area of the living cells in the trunks of ten trees of different species was measured, and the maximum pumping capacity at the most rapid pulsation rate found by Bose, was computed.

The actual rate of sap flow under maximum conditions of transpiration proved to be from eight to thirty thousand times as rapid as would be possible under Bose's theory, *Quercus rubra* giving the first value and *Robinia pseudacacia* the second, while the other eight species used presented an intermediate condition.

TRANSPORT OF SUBSTANCES ACROSS MEMBRANES

In a healthy cell there is a continuous interchange of water, ions, uncharged solutes, metabolites and dissolved gases across the plasmalemma. As one might expect, not all of these substances move through the membrane in the same manner. Firstly, some substances diffuse into a cell down a gradient of potentia energy; such movement is spontaneous and is the thermodynamic equivalent 0 heat passing from a warmer to a cooler body. There are, however, many substances which are accumulated by cells against a gradient of potential hence their movement into the cell is 'uphill' and is equivalent to the flow of heat from a cooler to a warmer body. 'Uphill' transport requires work to be performed and thus consumes energy. 'Downhill' transport is frequently described as passive while 'uphill' transport is described as active and directly involves the participation of cellular metabolism.

Thus, passive movements may occur by at least three types of pathway, whereas active movements must be linked to some energy-consuming mechanism, referred to as a 'pump', in the membrane. The third type of movement occurs because of the undulation and vesicularization of the membrane in endocytosis. Clearly this is a process which depends at some point on metabolism but, as is discussed below, the substances which move into the cell do not necessarily cross the membrane at all. In such circumstances the observed transport is not strictly *active* in a thermodynamic sense.

CRITERIA FOR ACTIVE ('UPHILL') TRANSPORT

To decide whether or not an ion or solute is actively transported across a membrane we need to know its activity or concentration in the two solutions separated by the membrane and for ions, in addition, we must know the electrical potential difference across the membrane. It is frequently difficult to measure the concentrations, and more difficult to measure the activity, of substances within cells with any accuracy, especially in those of higher plants.

In the giant cells of several sorts of algae, which may be 5,000 to 10,000 times the volume of a parenchyma cell in a root, such measurements are made routinely. The electrical potential difference across membranes can be measured if a small glass micro-electrode, with a tip diameter of 1 to 3*m* m, can be inserted into the membrane-bounded compartment.

Having made the necessary measurements a simple test can be applied to see if a given ion or solute within the compartment is at a higher or lower potential than in the surrounding solution. The principal snag in this analysis is that the cell or compartment should be in a steady state and that no net movement of solute should be occurring. In nature this condition is infrequently met.

THE NATURE AND ORIGIN OF THE MEMBRANE POTENTIAL

It is clear that electrical potential differences across membranes are of great importance in generating driving forces on ions. It is important, therefore, to try to understand how these potentials arise and how they are maintained.

An electric potential difference arises because positive and negative charges become separated. Since the cytoplasm of most cells is electrically negative relative to the surroundings it is very slightly enriched in anions relative to cations. This can be attributed to the differential permeability of the cell membrane and to the activity of ion pumps. First let us examine how differential permeability can create an electrical potential difference.

Diffusion Potential

Imagine a simple system of two compartments separated by a membrane which has a much higher permeability to K than to Cl. If the compartments are filled with potassium chloride solutions of different concentration, initially K will move through the membrane out of the more concentrated compartment, and for a very brief period, the more concentrated cell will lose K faster than Cl leaving it enriched in negative charge.

The negative *diffusion potential* thus created slows down the further escape of K by attracting it back into the more concentrated compartment. When the potential has developed, a large concentration difference can be maintained between the compartments. Since the membrane has a finite permeability to Cl′ albeit a low one, over a long period of time both the electrical potential difference (the diffusion potential) and the concentration difference would run down as Cl leaked through the membrane. If the membrane were completely impermeable to Cl′ the potential, once established, would be maintained indefinitely. In nature, membrane permeability to anions is a tenth to one hundredth of that for the monovalent cations; thus, the maintenance of a potential of the kind just considered depends on topping up the cell with anions at a rate comparable with their leakage into the surroundings. This is an 'uphill' transport and therefore requires the mediation of some ion pumping mechanism. Active transport is, therefore, necessary to maintain a diffusion potential.

MEMBRANE PUMPS

Pumping mechanisms which allow cells to accumulate solutes up gradients of potential can contribute to the membrane potential in both an indirect and a direct way. The former type are referred to as neutral exchange pumps in which the cell dumps an unwanted ion of equal and like charge into the external environment in a one-to-one exchange for an ion which is more useful, *e.g.* there are well known exchanges of cellular Na or H for K from the surroundings.

The second type transports an ion in one direction only without coupled exchange and is known as *electrogenic* since charge is separated. They can, therefore add to, or subtract from the diffusion potential, described above, depending on which ion is carried.

Neutral Ion Pumps

Neutral ion pumps are present in the plasmalemma of all cells and by their activity they create the ionic asymmetry necessary to set up the diffusion potential described above. As a much simplified illustration of this, consider a cell which, because of its synthetic and respiratory activity, is generating H and HCO_3 internally. A pair of exchange pumps could swap H and HCO_3 for K and Cl from the surroundings quickly enriching the interior in these ions and thus setting up the conditions in which the differential rates of diffusion of K and Cl out of the cell give rise to a membrane potential. But how does such a pump actually work? There are fewer detailed examples than one would like but the best known is of the membrane-bound ATPase which exchanges intracellular Na for extracellular K in many animal and plant cells. In the red blood cell it is known that Na is one of the cofactors which is essential for the binding of ATP to the ATP-ase enzyme. *In vivo* the active centre of the enzyme is accessible only from the cytoplasm, so that the ATP and the Na must be inside the cell.

Once bound, the ATP is hydrolysed, ADP is released into the cytoplasm leaving the cleaved terminal phosphorus atom attached to the active centre to form a phosphoenzyme. These reactions result in some molecular re-orientation of the phosphoenzyme and its attached Na which exposes the ion-binding site to the different chemical environment of the external medium. It is proposed that this change of environment alters the ion-specificity of the binding site so that K is favoured; K thus replaces Na. This done, there is a second re-orientation (referred to as 'relaxation') which carries the bound K to the inside. The phosphorus is released from the active centre and Na' which is preferentially bound on the cytoplasmic side exchanges for K and the pump is ready for a second cycle. The pump has used the free energy released on hydrolysis of ATP as fuel to exchange K and Na against their respective electrochemical potential gradients. The two ions in the appropriate orientations are essential cofactors in the enzyme reaction; *in vitro,* ATPase of this kind will not hydrolyse ATP unless Na and K are both present. In theory many pumps based on ATPase are possible with only subtle modifications of the ATPase molecule to provide binding sites of varying field strength which will select various ions, *e.g.* a Ca transporting ATPase is found in mitochondria and in sarcoplasmic reticulum (Racker, 1972).

Electrogenic Pumps

The unidirectional transport of an ion across a membrane separates charges and in so doing provides a driving force for the passive diffusion of a similarly charged ion in the opposite direction or an oppositely charged ion in the same direction. The molecular details of exactly how an electrogenic pump is put together remain uncertain although in one instance it is highly likely that an electrogenic H -efflux pump is based on an ATPase (Slayman *et al.*, 1973). It is possible, nevertheless, to deduce certain general consequences of their operation. If, for instance, there was an outwardly directed pump at the plasmalemma which actively pumped hydrogen ions (protons) out of the cell thus making the interior electrically negative, this could contribute to the electrical driving force on the diffusion of K from the external medium. Indeed the rate at which charge is extruded and the rate at which it leaks back into the cell must be very nearly in balance unless a dangerously large potential is to accumulate. Examples of both proton extrusion pumps and anion influx pumps of the electrogenic kind are well documented from research on plant tissues.

In the giant alga, *Acetabularia,* an electrogenic chloride influx pump contributes more than half of the potential of –170mV found across the plasmalemma when the cell is kept in the light. Almost immediately the cell is put in the dark the pump stops working (since it is closely linked with photosynthesis) and the membrane potential abruptly depolarizes to –80mV (Saddler, 1970). A similar light-dependent electrogenic pump is found in *Nitella translucens* (Spanswick, 1972, 1974). Electrogenic pumps are not, however, restricted to green tissues but have been reported in plant roots (Higinbotham *et al.*, 1970) and fungal hyphae (Slayman, 1970). In every instance, however, inhibition of the pump caused an immediate depolarization of the membrane potential, indeed this is often used to detect the activity of such a pump. The inhibition of a neutral ion pump gives rise to gradual depolarization as the ionic asymmetry runs down.

MEMBRANE CARRIERS

Many ions and uncharged solutes cross membranes more rapidly than could be expected if they passed through the membrane lipids without some assistance. Since pores in membranes are too narrow to accommodate many substances which are transported, it is widely believed that membranes contain carrier molecules which, in combination with the solute, facilitate diffusion.

The ion pumps considered above are carriers of a special kind since they are linked to the metabolic activity of the cell; the carriers we shall now consider promote net movements of solutes only down the prevailing potential gradient and are not capable of 'uphill' transport even if in some instances they appear to be doing so.

Evidence from Kinetics

One widely used approach to gather information about carriers has been the application of kinetics first derived from enzyme reactions. The evidence for these carriers is obtained by placing a cell or tissue in a range of solute concentrations and measuring the initial rate of uptake. The uptake rate shows a tendency to saturate at higher concentrations and can thus be used to calculate the maximum velocity, V_{max}, possible under the conditions used in the experiment.

By making a double reciprocal plot of the data the concentration of solute at which half maximal velocity is achieved can be estimated. This is known as the Michaelis Constant, K_m. The K_m measures the affinity of the carrier for the solute it carries; if the affinity is high then the concentration, K_m will be low and *vice versa.*

For most ions in plant tissue K_m is quite low, concentrations ranging from 5–100 *m* M, but for sugars and other metabolites K_m values are usually greater than 300 *m* M. Much work of this kind is summarized by Epstein (1972) who shows that at concentrations less than 0.1 mM, the uptake of a given ion is not subject to serious interference from other common ions in solution. There is, however, competitive inhibition between related ions of similar molecular dimensions, *e.g.* K uptake is inhibited competitively by Rb but not by Na ; Ca is inhibited by Sr but not by Mg.

Thus the carriers which bind the major nutrient ions at low concentrations appear to be highly ion-selective. At higher concentrations (more than 1.0–10.0 mM) this selectivity begins to decline.

The interpretation of this observation is contentious and beyond the scope of this chapter but can be pursued in Epstein (1972), Laties (1969) and Clarkson (1974).

The limitation of the kinetic approach is that it can tell us nothing about the nature of the carrier. One can observe similar uptake kinetics for ions whose transport into the cell must be mediated by ion pumps *e.g.* H_2PO_4 and Cl as for ions which probably diffuse into the cell passively *e.g.* Na and Ca and for those which are completely exotic and toxic, *e.g.* Tl (Barber, 1974). Indeed, it has been pointed out that saturation kinetics of this kind would also be found if salt movement was observed across a synthetic membrane containing nothing but pores (Stein & Danielli, 1956), where the system would saturate when all of the pores were filled with solute at any moment in time; V_{max} is, after all, merely a measurement of capacity to react or transport and K_m is derived from it.

Ionophores as Lipophilic Carriers

A more illuminating approach to the nature of carriers has come from studies on the ionic conductance of synthetic membranes which have been modified in various ways. A bilayer of pure phospholipids has a very low conductance to ions, usually only 10 to 10 ohm cm. The addition of very small amounts of ionophores (*i.e.* ion-carrying antibiotics) like monactin or valinomycin to the solutions bathing the synthetic bilayer causes a huge increase in the conductance. 10 M monactin changes the membrane conductance to K nearly a million-fold and that even at 10 M its effect is quite strong.

The conductance change for a given monactin concentration is greatest for K and for Rb and is much less for Na' Cs and Li. Monactin is, therefore, acting as a selective carrier of K and Rb and valinomycin, another bacterial product, behaves similarly. Since these two compounds differ chemically it is instructive to see what they have in common. Both of them are amphipathic ring-structured molecules which have their non-polar groups on the outside of the ring and their polar groups directed towards the space at the centre of the molecule. The outside of the ring interacts favourably with lipid while the hydrophilic core, 0.7 nm in diameter, provides room for several hydrated potassium ions to be bound. Evidence from a variety of sources shows that this complex diffuses across the membrane so that the ions never leave a polar environment (Eisenman *et al.*, 1968). Other substances are known which select for divalent cations, *e.g.* the unnamed compound A23187 which is a carboxylic acid antibiotic found in cultures of *Streptomyces chartreusensis* (Reed & Lardy, 1972). This compound carries Ca and Mg across bilayers and natural membranes but has no effect on monovalent cations. Apart from a few synthetic analogues all of the ion-carrying antibiotics are natural products of bacteria and fungi. There are many who believe that compounds of a similar kind may act as ion carriers in all membranes but the technical difficulty of isolating what are probably minute quantities of such compounds from tissues appears to be formidable and so the belief may rest on faith for some time yet.

Co-transport

As suggested earlier, carrier-assisted diffusion can sometimes appear to go in an 'uphill' direction, thus giving the impression of active transport. In many animal tissues and micro-organisms, sugars, amino acids, organic acids and vitamins move into the cell up a concentration gradient. This transport is, however, almost completely dependent on having Na or H in the external medium; other ions such as K' Rb or Li cannot be substituted. It has been found that the metabolite is carried into the cell along with an ion-carrier complex which is diffusing 'downhill'. In both *Chlorella* and *Neurospora,* glucose is transported in this way along with protons, H (Komor & Tanner, 1974; Slayman & Slayman, 1974).

Co-transport depends on active transport in an indirect way (as indeed does all diffusion) because energy-dependent extrusion pumps ensure that the cytoplasm is kept well below its equilibrium concentration in H and Na. These ions tend to diffuse back into the cell and, in doing so, *decrease* their free energy. The energy they give up is coupled, *via* the carrier, to the co-transport of the solute whose free energy is *increased* as it moves into the cell.

Co-transport may also assist the 'uphill' movement of inorganic ions into the cell; it may be a more common process than is generally realized. Recent evidence by Lowendorf *et al.*, (1974) suggests that the active transport of phosphate into the hyphae of *Neurospora* depends on (a) the activity of a proton extrusion pump at the plasmalemma which is sensitive to the pH of the external medium, and (b) the formation of a ternary complex between a proton and a phosphate ion from the external medium with a membrane carrier. The protonated phosphate carrier diffuses to the cytoplasmic side of the membrane *down* the electrochemical gradient of the proton, releasing the proton and phosphate ion into the cytoplasm. This, and other examples of inorganic ion co-transport suggest a reason why so little progress has been made in elucidating the molecular details of certain 'pumps' particularly of those which transport anions. Put most simply, it may be that these influx pumps do not exist and that the uphill transport is driven by a combination of active extrusion and re-entry by diffusion of protons or perhaps sodium ions.

Co-transport illustrates the ingenious way in which nature can turn necessity to its own advantage. The active excretion of H and Na is essential to maintain pH control and osmo-regulation in the cell, but the energy expended is partly recovered in the transport of essential metabolites and ions into the cell.

MEMBRANE PORES AND CHANNELS

Lipid or oil is a very effective electrical insulator (it is used in high voltage underground electric cables for this purpose) and it is not surprising to find that the electrical resistance of synthetic bilayers made from pure phospholipid is very high, being 10 to 10 ohms cm. In nature, the current conducted across membranes is carried by ions and we can see that an unmodified lipid bilayer with such a high resistance is a poor material across which to conduct this essential current of electrolyte. It is not surprising to find, therefore that the electrical resistance of natural membranes is very much less than that of bilayers, usually lying in a wide range of 10 to 10 ohms cm.

As described lipophilic carriers can greatly reduce the resistance (and hence increase the c onductance) of synthetic membranes and may resemble the carrier molecules in membranes, but it is also widely believed that membranes contain water-filled pores which must contribute to their relatively high conductance by allowing water and selected solutes to by-pass the lipid domain of the membrane. Certain antibiotic molecules, such as nystatin, appear to condense cholesterol molecules in both synthetic (Holz & Finkelstein, 1970) and natural membranes (de Kruijff & Demel, 1974) to form pores with a radius of *ca.* 0.4 nm. The presence of such pores greatly increases the electrical conductance and hydraulic conductivity of the membrane. In passing we might note that many antibiotics have their effect by enormously increasing the passive permeability of cell membranes causing non-resistant cells to lose their contents or to lyse. Interest in these substances stems from the experimental evidence that cell membranes also possess pores of similar size and that the antibiotic merely induces an extreme expresssion of the normal condition.

Although the word 'pore' is often used to describe channels through which solutes and water can move, we should resist the temptation to conclude that all 'pores' are definable structural entities like the ones induced by nystatin. In many instances a 'pore' may be more like a transient imperfection in membrane structure. This latter type may have a certain statistical probability but have no fixed position.

Proteins which traverse the lipid layer may give rise to hydrophilic channels or pores. Indirect evidence supporting this idea comes from a study in which red blood cell membranes were exposed by deeply etching frozen cells under vacuum (Pinto da Silva, 1973). Shrinkage of the membrane surface was observed in areas overlying groups of membrane particles possibly due to sublimation of ice from within the embedded particles; for this to have occurred this water would be a free liquid in the thawed condition.

The first circumstantial evidence for the existence of pores come from studies by Collander and Barlund (1933) on the permeation of the giant internodal cells of the alga, *Chara ceratophylla,* by a number of uncharged solutes and water. They found that in almost every case the rate of movement of a substance into a cell depended on its molecular weight and dimensions and on its solubility in oil relative to water; substances with high oil solubility permeated most rapidly. This strongly suggested that movement across cell membranes involved the movement of the solute out of the water, its solution in lipid and subsequent diffusion through it, and its re-entry into the aqueous phase at the inside face of the membrane. The authors found, however, that several small molecules permeated the membrane far more rapidly (more than 100 times faster in the case of water) than their relative solubility in oil suggested. The upper size limit for molecules which behaved anomalously was a radius of 0.4 nm and it was suggested that the membrane was constructed as a very fine sieve containing pores of 0.4 nm radius through which water (0.25 nm radius) and certain solutes could move. Since this early work a great deal has been done and the equivalent pore radius in many plasma membranes has been confirmed as 0.4 nm. It must be said, however, that some authorities are reluctant to accept that pores can provide channels for the bulk movement of water and solutes; the arguments for and against pores have been clearly discussed in Oschman *et al.,* (1974).

There has been much discussion about whether pores admit ions, and if so, which ones. Membranes are known to control very precisely the relative rates at which various ions will diffuse across them; potassium ions will diffuse ten to one-hundred times faster than sodium. Ions in solution are hydrated by binding one or more water molecules; it requires a great deal of energy to dehydrate an ion and for this reason we should think of ions in all natural circumstances as being hydrated. Sodium binds 5 water molecules, whereas potassium binds 3, the former is, therefore, the more bulky ion whose diffusion into the narrow water filled pores would be slower than for the smaller hydrated potassium ion. On the other hand the anion chloride, which has only one water molecule in its hydration shell, diffuses much more slowly than either K or Na. This is probably due to the fact that 'pores' carry a predominantly negative charge so that cations would be attracted to them, while anions would be repelled—a small number of positively charged pores would handle the flow of anions. Polyvalent cations and anions have much more water in their hydration shells, *e.g.* Ca has 10 and SO_4 has 8 and it seems likely that these would be totally excluded from the pores.

ION PUMPS AND MEMBRANE SUBSTRUCTURE

It is most unlikely that the intercalated membrane particles described earlier are simple ion-carriers—they are much too large—but they may be ion pumps, or groups of pumps. The best evidence for this assertion comes from work on the ATPase which pumps Ca across the membranes of sarcoplasmic reticulum. Racker (1972) was able to isolate this enzyme and put it back into a synthetic bilayer membrane made from soybean phospholipids, and thus reconstitute a membrane which could actively transport Ca when ATP and Mg were present in the medium. Working with a slightly different system Packer *et al.,* (1974) showed that freeze etched fractures of both the original tissue membrane and reconstituted membrane containing the purified ATPase were densely studded by numerous particles 8.5 nm in diameter. There can be little doubt that the particles were the calcium-pumping ATPase. Similar, less complete, evidence is also available for the Na /K -dependent ATPase of the red blood cell. There is no insuperable difficulty in repeating such observations with plant tissues but, at the time of writing there is no report that this has been done.

One should not conclude that all of the membrane substructure is concerned with solute transport, especially in mitochondria and chloroplasts where other biochemical activities are centred on membranes.

ENDOCYTOSIS AND VESICULAR TRANSPORT

The presence of abundant vesicles in the vicinity of the plasmalemma of plant cells has been pointed out earlier (p. 44); however, it is not yet clear what contribution endocytosis makes to the total solute transport across this, or any other membrane. There is kinetic evidence from several sources which is consistent with the notion that a measurable fraction of various ions in the cytoplasm of plant cells is sequestered into a compartment separate from the bulk of the cytoplasm. MacRobbie (1969) found that labelled Cl transported into the vacuole of *Nitella translucens* could be easily resolved into a fast and a slow component. The fast component was envisaged as being delivered to the vacuole formed in vesicles which had been found at the plasmalemma. The chloride ions in these vesicles would not have mixed with the unlabelled chloride in the bulk cytoplasm, whereas labelled Cl delivered directly to the cytoplasm *via a* pumping mechanism would have to mix with a much larger pool of unlabelled chloride ions. At the beginning of the experiment, therefore, the labelled Cl in the vacuole increased more rapidly than expected. There are other examples of this kind discussed in MacRobbie (1971) and Baker and Hall (1973).

If the vesicles contained nothing more than a small volume of the outside solution, their net contribution to the solute content of the cell would be very small and they could not exercise the ion-selectivity which characterizes cell membranes. Baker and Hall (1973) suggest that endocytosis becomes a much more plausible transport process if it is assumed that ions become selectively bound to the membrane surface and thus become concentrated from the dilute external medium.

This assumption is entirely reasonable because both proteins and phospholipid heads can provide ion-selective binding sites. These authors also relate membrane-bound ATPase activity with the subsequent invagination of the plasmalemma to form a vesicle but this implies that vesicle formation is energy-dependent. This is against the substantial evidence from animal cells which shows that micro-vesicle formation results from Brownian movement of the membrane and does not need to be energized by ATP (Casely-Smith, 1969); ATP is required for the very large vesicles formed by re-orientation of micro- fibrils in phagocytosis, but the type of vesicle under discussion here is much smaller, probably 70–100 nm in diameter and is not dependent on microfibrils. There is visual evidence that these micro-vesicles can discharge their contents directly into lysosomal vacuoles (Casely-Smith & Chin, 1971).

The important advance of thinking required to deal with endocytosis will provide a severe strain on biophysical theorists accustomed to thinking of cells as homogeneous phases separated by diffusion barriers, partly because materials can reach the interior of the cell without crossing a membrane at all. Although endocytosis is probably a common phenomenon it is likely to remain an 'unknown quantity' until new experimental techniques are devised.

5

The Mechanics of Plant Transpiration

FOLIAR TRANSPIRATION

The subsequent discussionwill deal almost entirely with transpiration from leaves, since, in most plants,the amount of water-vapour lost from other organs is comparatively small. The mechanics of foliar transpiration can be adequately discussed only inreference to the anatomy of the leaves from which it occurs.

It should be recalled that the vacuoles of all of the living cells of a leaf are filled with water, which also saturates the protoplasm and the cell walls, this water being supplied to the leaf cells through the water conducting tissues of the vascular bundles.

Hence evapouration of water will occur from these wet cell walls into the internal atmosphere of the intercellular spaces just as it will occur from any wet surface into the surrounding air. The intercellular spaces constitute a connected system, ramifying throughout the leaf.

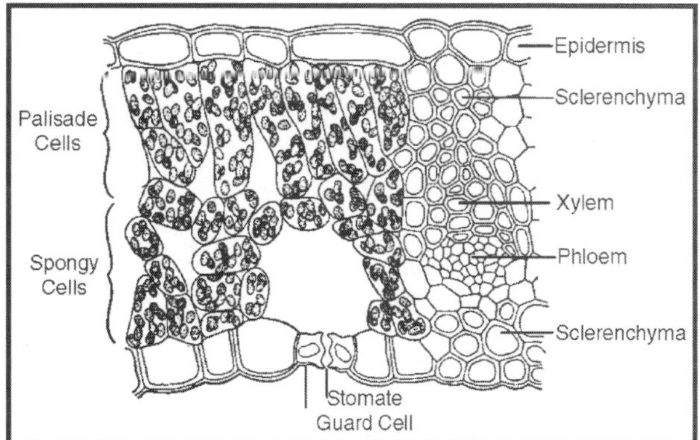

Fig. Cross Section of a Portion of a Leaf of Tulip Tree.

If the stomates are closed the only effect of evapouration from the mesophyll cell walls will be the saturation of the entire volume of the intercellular spaces with water-vapour.

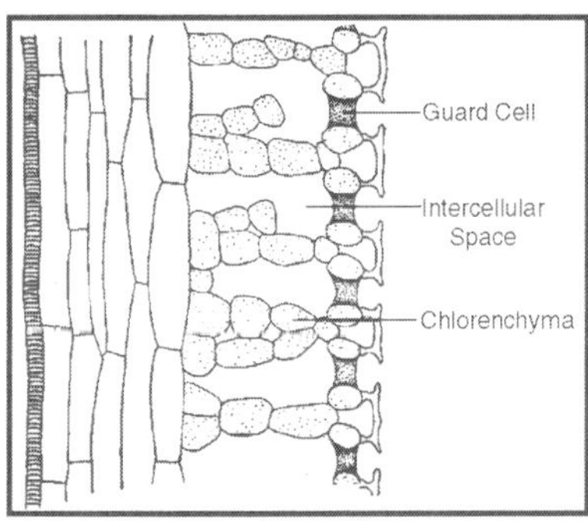

Fig. Longitudinal Section(Semi-diagrammatric) Through a Small Portion of a Leaf of White Pine(Pinus Strobus).

When the stomates are open, however, diffusion of water-vapour may occur through them into the outside atmosphere. Such outward diffusion will always take place unless the atmosphere has a vapour pressure equal to or greater than that of the small portion of a leaf which does not commonly exist during the daylight hours. The rate of such diffusion will depend principally upon the excess of the vapour pressure in the leaf over that of the atmosphere, although the diffusive capacity of the stomates is also an important factor.

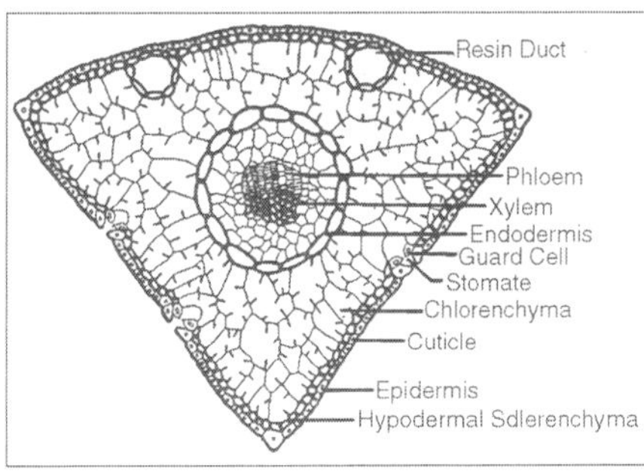

Fig. Cross Section of a Leaf of White Pine

The process of stomatal transpiration therefore involves evapouration from the cell wall surfaces bounding the inter cellular spaces and the diffusion of this water-vapour from the intercellular spaces into the atmosphere through the stomates. One side of every epidermal cell on a leaf is also exposed to the atmosphere. Evapouration of water occurs into the atmosphere directly from these cell surfaces. The surfaces of practically all aerial leaves are covered with a layer of wax-like substance known as cutin reduces transpiration directly through the walls of the epidermal cells to a magnitude much less than it would be, were there no such layer present.

The thickness of the cutin layer varies with the species of the plant, and the environmental conditions under which the leaves have developed. The layer of cutin is usually thicker on leaves which have developed in bright sunlight for example, than on leaves of the same species which have developed in shade. Even in leaves which are heavily coated with cutin, some cuticular transpiration occurs, possibly largely through tiny rifts in the cutin layer. In most species of plants of the temperate zone less than 10 per cent of the foliar transpiration occurs through the cuticle, the remainder being stomatal transpiration.

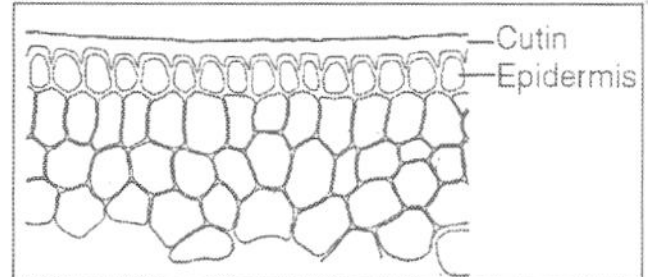

Fig. Citin Layer on the Upper Epidermis of a Leaf of Clivia Nobilis.

THE STOMATAL MECHANISM

The most important physiological fact about the stomates is that they are sometimes open and sometimes closed. When open they serve as the principal pathways through which gaseous exchanges take place between the intercellular spaces of the leaf and the surrounding atmosphere. When closed all gaseous exchanges between a leaf and its environment are greatly retarded.

The gases of greatest physiological importance which enter or depart from a leaf principally through the stomates are oxygen, carbon dioxide and water vapour. The movement of gases through the stomates in either direction is primarily a diffusion phenomenon, although under certain conditions, as for example when the intercellular spaces are alternately compressed and expanded by bending of the leaf in a high wind, alternate outward and inward mass movements of gases may occur.

Although the stomates are the principal portals through which entry and escape of gases takes place, the fact shouldn't be overlooked that at least small amounts of these three gases pass directly through the epidermis and cuticular layers of all leaves. In submerged aquatics all gaseous exchanges between the plant and its environment occur through the epidermis.

STRUCTURE OF THE STOMATES

The stomates or stomata (singular stomate or stoma) are minute pores which occur in the epidermis of plants. They are surrounded by two specialized epidermal cells known as the guard cells. Stomates may occur on any part of a plant except the roots, but in most species are most abundant upon the leaves. The size of the stomatal pore varies in most plants depending upon the turgidity of the guard cells and often, especially at night, it is entirely closed. The structure of stomatal apparatus shows marked variations in detail indifferent species of plants, but the essential feature of a pre between two modified epidermal cells is common to all species of vascular plants. Guard cells which are roughly kidney or bean-shaped as seen in surface view are typical of more species of plants than any other kind. In some species the epidermal cells bordering on the guard cells are different in con figuration from other cells in the same tissue; these are called subsidiary cells or accessory cells. In many species of the grass and sedge families the guard cells are distinctly elongate.

In most species of conifers, and in certain other species, stomates are of the sunken type The guard cells differ from the other epidermal cells in that they contain plastids which are green in colour. These have often been assumed to bechloroplasts, and are frequently so-called, but the green pigment present does not seem to be true chlorophyll as photosynthesis apparently does not occurring the guard cells.

SIZE AND DISTRIBUTION OF STOMATES

The size of the stomatal pore varies greatly according to the species of plant, and somewhat even among the individual stomates on any one plant. They are always very minute however, their dimensions being expressed in terms of microns. Minute as these openings appear to be from a human scale of values they are enormous when compared with the size of the gas molecules which diffuse through them. The calculated diameter of a water molecule is 0.000454. More than 2,000 water molecules would have to be placed side by side to measure a distance of 1.

The molecules of both carbon dioxide and oxygen are larger than water molecules. Since the stomatal diameters usually are considerably in excess of 1, it is evident that the stomates afford relatively enormous portals to the gas molecules which diffuse through them. In general the number of stomates present in the epidermis of leaves may range from a few thousand to over a hundred thousand per square centimeter the exact number depending upon the species and upon the environmental conditions under which the leaf has developed. A single maize plant has been estimated to bear from 140 to 240 million stomates, while the number on a large tree could be expressed only by a figure of astronomical dimensions. However, marked deviations from such average values are possible for any species, depending upon the environmental conditions under whichthe leaves have developed. The number of stomates per unit area of leaf surface may be quite different on leaves of two plants of the same species if one grew in a greenhouse, and the other grew in the open, or upon the leaves of plants of the same species which have developed during different seasons.

Stomates occur in both the upper and lower epidermis of many species of plants. In numerous others, especially woody species, they are confined to the lower epidermis. Even in those species in which stomates occur on both surfaces of the leaf they are commonly, but not always, more abundant in the lower epidermis. In floating leaves such as those of the water-lily, stomates occur only in the upper epidermis.

Species in which the stomates are relatively small usually have more per unit area than species in which the stomates are relatively large. The number of stomates per unit area usually varies on different leaves of the same plant, and even in different parts of the same leaf. On individual leaves it appears to be generally true that the greatest number of stomates per unit area is at the tip, the lowest towards the base, while the middle portion of a leaf shows a frequency midway between these two extremes. As a rule, the higher the point of attachment of a leaf on the stem of a plant the greater the number of stomates per unit area. In general no correlation has been found between transpiration rates and either the size or distribution of stomates, other factors being much more important in determining the rate of loss of water-vapour from the intercellular spaces.

PRINCIPLES GOVERNING DIFFUSION OF GASES THROUGH THE STOMATES

Since virtually all gaseous exchanges between the intercellular spaces and the atmosphere take place through the stomates, the problem of the diffusive capacity of the stomata1 pores is an important one. Although the aggregatearea of the fully open stomates on a single leaf is seldom more than 3 per cent of the leaf area and is often as low as i per cent, the rate of water loss from the leaves (*i.e.* loss per unit area in a unit time) in many species maybe, under favourable conditions, as much as 50 per cent of the evapouration from an exposed water surface of comparable dimensions. Stlfelt (1932) records that transpiration from a leaf of birch (Betula pubescens) under the most favourable conditions may be 65 per cent of the evapouration from a comparable evapourating surface, but this is probably an unusually high value. It is therefore evident that water-vapour often diffuses through the stomates at rates ranging up to at least 50 times greater than it diffuses away from an equal area of exposed evapourating surface.

This unexpectedly high diffusive capacity of the stomates is intelligible in terms of the results of studies upon the principles of diffusion through small apertures. The classical investigation of this problem was made by Brown and Escombe who studied the rate of diffusion of carbon dioxide through tubes of known dimensions. They first made the important discovery that if a septum which had been pierced with a small circular aperture was interposed across a column of diffusing gas, that the rate at which carbon dioxide diffused through this aperture was much greater than the rate at which it passed through an equal area of the unobstructed tube.

Since the problem at present under consideration is the diffusion of water-vapour rather than carbon dioxide through small apertures will be used to illustrate more fully the principles regarding the diffusion of gases through small openings. These data were obtained by sealing thin septa, through the centre of which were cut round openings of known dimensions, across the circular mouths of small bottles which had previously been filled to a certain level with water, and then determining the loss in weight of each bottle after all of them had been kept under uniform conditions for the same period of time.

The results of this experiment illustrate two important general principles:

1. The quantities of water-vapour diffusing through small openings in a given period of time are propor-tional (essentially) to the perimeters (circumfe-rences) and not to the areas of the pores.
2. The smaller the pore, the greater the water loss per unit area. The pore in septum 2, for example, has an exposed area only slightly more than one-third as great as the pore in septum I, but diffusion through it is nearly two-thirds as great as through septum i. Similarly septum 10, with only half the exposed area of septum 9, permits almost as much diffusion to occur as through sputum 9. These results indicate that the rate of diffusion of water-vapour through a small aperture is greater than through an equal area of a largere vapourating surface.

In other experiments, Sayre found that diffusion of water-vapour through small openings of elliptical shape is also more nearly proportional to their perimeters than to their areas.Hence the perimeter law for the diffusion of gases also holds for openings of other than circular shape. An important inference which can be drawn from this finding is that a stomate attains almost its maximum diffusive capacity long before it is fully open, since the perimeter of a stomatal pore does not increase greatly after the aperture between the guard cells has widened perceptibly.

The fact that the diffusion of gases through small apertures is much more nearly proportional to their perimeters than to their areas is due to the more rapid diffusion of the molecules through an opening at the peripherythan through its centre. A molecule of the diffusing gas at the rim of the hole is surrounded on all sides by other molecules moving in all directions.The concentration of diffusing molecules is much less in an outward direction from the rim, however, than towards the centre of the opening. Furthermore, the diffusion gradient is much steeper from the rim outwards than at points above the opening. Hence the number of molecules escaping through the hole from a given point near the rim per unit time interval will greatly exceed the number escaping from a point near the centre of the opening. Reduction in the area of any circular or elliptical opening results in increasing its perimeter relative to its area. Hence in small apertures such alarge proportion of the diffusion occurs at the edge of the opening that the effect of any diffusion which occurs through its centre is almost if not completely obscured, and measurements show the rate of diffusion through the small apertures to be essentially proportional to their perimeters. In fact in an opening 0.5 cm. in diameter, for example, a large part of the centre of the aperture can be entirely blocked off without any very great effect upon the rate of diffusion of a gas through it.

That the diffusion of gases through small apertures occurs largely at the rim of the hole can be demonstrated visually by means of a simple experiment. A cylindrical glass vessel is partly filled with a gelatin sol which is allowed to gel. A small opening is punched through the centre of a disk of celluloid which is then inserted into the position on top of the gelatin,the gap between the edge of the disk and the wall of the cylinder being sealed with melted vaseline in order to prevent leaks. A solution of potassiumpermanganate, which is purple in colour, is now introduced into the vessel ontop of the disk.

After a few hours the molecules of potassiumper-manganate, after diffusing through the aperture, become distributed in such a way as to occupy a hemispherical zone below the opening. The hemispherical distribution assumed by molecules which are diffusing through a small opening is often spoken of as a diffusion shell. Gases also form diffusion shells above small pores but their configuration is often relatively unstable due to their ready distortion by wind or convection currents. Actually when molecules are diffusing through a small orifice there is formed not one, but two diffusion shells, one on each side of the diaphragm which is pierced with the opening.

The experiment described above permits discernment of the hemispherical pattern of diffusion only on the receiving side of the system.There is also a second shell, a mirror image of the first adjacent to the aperture on the side of the system from which diffusion is occurring. The deep colour of the permanganate solution prevents visible detection of this second shell in the experiment described above.Diffusion of gases from or into a leaf through the stomates involves a much more complex system than is represented by a septum pierced by a single aperture. Diffusion is occurring not through a single opening of potassium permanganate upon diffusion. Experiments into gelatin have been performed in analogous physical systems in which multiperforate instead of uniperforate diaphragms have been used. The diffusion per pore increases with increase in the distance apart of the apertures, although not proportionately. With pores of this diameter (0.3 mm.) nearly the maximum diffusive capacity is attained when they are spaced at intervals of 20 diameters. The closer together the pores in a septum the greater the overlapping of the diffusion shells. The molecules diffusing through each aperture invade in part the zones into whichthe molecules passing through neighbouring pores would diffuse were each opening the only one in the septum. Hence the diffusion gradients are less steep than they would be were there only one pore in the septum.

As a result the rate of diffusion through each pore is less than it would be through a pore of equal dimensions in a uniperforate septum. The loss of water-vapour by diffusion per septum decreases with increase in the distance between the openings, *i.e.* with decrease in the number ofpores per septum. The decrease in diffusive capacity is not however, in proportion to the reduction in the aggregate area of the pores in the septum. For example, when the pores are spaced 5 diameters apart their aggregate area is only 3.38 per cent of the septum area, yet diffusion through them was 62 per cent of that through an open bottle with a mouth of the same area as the septum. The dimensions of the pores (0.3 mm. diameter, 0.0707 mm area) used in the experiments just discussed are much greater than those of the average stomate.What about the diffusive capacity of still smaller pores? Huber has shown that the smaller and more numerous the pores, aggregate pore area being constant., the greater the diffusion per septum.1 For example,when pores 0.05 mm. in diameter were used, occupying 3.2 per cent of the1 If equal-sized pores are spaced equidistantly, and represent the same proportionate area of the septum, they will be the same distance apart in terms of their diameters, regardless of pore size. Hubers results indicate that the smaller the pores for a given aggregate pore area the more nearly the diffusion through the septum will approach that of a free evapourating surface. In the sunflower and undoubtedly also in some other species the aggregate area of the stomatal pores is about 3 per cent of the leaf area.

If diffusion through pores 0.05 mm. in diameter,occupying 3.2 per cent of the area of the septum, is 72 per cent of that from a free evapourating surface, diffusion through the much smaller stomates of this species should be proportionately greater. In such species it seems probable that the aggregate diffusive capacity of the stomates mayclosely approach that of a free evapourating surface. Even in species in which the total area of the stomatal pores is less than in the sunflower their aggregate diffusive capacity is very large relative to their area. The relatively high rates of water loss which occur from leaves in proportion to the area of the stomatal pores are thus explicable in terms of the principles of diffusion through multiperforate septa.

The theoretical diffusive capacity of the stomates is more than adequate to account for known rates of transpiration. This is shown by the calculations of Brown and Escombe for sunflower leaves. Assuming a leaf temperature of 200 C., the intercellular spaces to be saturated with water vapour at that temperature (vapour pressure 17.54 mm. Hg), and the vapour pressure of the atmosphere to be one-fourth of this value, their computations show that transpiration would occur at the rate of 17 g. of water-vapour per square decimeter of leaf surface per hour. This is several times greater thanthe maximum rate of transpiration ever recorded for a sunflower plant.

Evidently some other factor than the diffusive capacity of the stomates has a limiting effect upon transpiration whenever the stomates are widely open. The foregoing discussion of the diffusion of gases through small openings has been based on the assumption that diffusion is occurring into a quiet atmosphere. If an air current is blowing across the surface of a multiperforate septum with a sufficient velocity to prevent the formation of diffusion shells the result is a steeper diffusion gradient and hence a greater rate of diffusion of gas through the pores of the septum. Hence, when subjected to air movements, the diffusion of water through multipe-rforate septa or the stomates may be even greater than the preceding discussion has indicated. Within limits increase in the velocity of the wind results in a progressive increase in the rate of diffusion through a multiperforate septum.

THE MECHANISM OF THE OPENING AND CLOSING OF THE STOMATES

The degree of stomatal opening is influenced both by changes in the turgor of the guard cells and by changes in the turgor of the epidermal cells, although the former usually play a predominant role. In general an increase in the turgor of the guard cells relative to that of the epidermal cells leads to a widening of the stomatal aperture, and vice versa. The greater this turgor difference, the wider the stomatal aperture. The mechanism of the effect of changes in the turgor of the guard cells upon the size of the stomatal aperture varies with the structure, form, and position of the stomates. In one type of guard cell, found in many different species of plants, the cell wall is thicker on the side bordering the stomatal pore than on the side bordering the epidermal cells.

With an increase in turgor the thinner walls of the guard cells are stretched more than the thicker; this causes the thicker-walled sides to assume a concave shape and results in the appearance of a gap the stomatal pore between the two guard cells. Opening of the stomates typical of the grass and sedge families appears to be due to swelling of the ends of the guard cells thus separating the abutting walls of the middle portion of the two adjacent guard cells. In the sunken stomates typical of conifers opening of the stomates seems to result largely from a change in the shape of the guard cells due to an increased turgor which is unaccompanied by any appreciable stretching of the walls. These various types of stomatal mechanisms are discussed by Copeland.

FACTORS AFFECTING TRANSPIRATION

The rate of transpiration of a plant or any leaf on a plant varies from day to day, from hour to hour, and, frequently, from minute to minute. Variations in the rapidity with which water is lost by plants are due to the effects of environmental factors upon physiological conditions within the plant.

The important environmental factors influencing the rate of transpiration are:

- Solar radiation,
- Humidity,
- Temperature,
- Wind,
- Soil conditions influencing the availability of water,
- Atmospheric pressure.

This last factor is relatively much less important than the other five listed.While the general effect of variations in the intensity or magnitude of each of these factors upon transp-iration is well known, and has frequently been demonstrated by experimentation, the precise mechanism of the effect of each is not so easily amenable to experimental treatment. The following interpretation of the mechanism of the effects of these environmental factors upon the rate of transpiration is therefore a somewhat theoretical one, but is in accord with the experimental data available at the present time.

SOLAR RADIATION

This term refers to the visible light and other forms of radiant energy (infrared and ultraviolet radiations) reaching the earth from the sun. The principal effects of solar radiation upon transpiration result from the influence of light upon the opening and closing of the stomates. In most of the species of plants which have been studied the stomates are usually closed in the absence of light, thus causing a virtually complete cessation of stomatal transpiration during the hours of darkness.

Since none of the other environmental factors can have any influence upon stomatal transpiration except when the stomates are open, light occupies a position of prime importance among the environmental conditions influencing transpiration. A second important effect of solar radiation upon transpiration operates through its influence on leaf temperatures. In direct sunlight the temperature of leaves is almost invariably higher than the air temperature. This effect will be analysed later in this chapter.

HUMIDITY

Several units are used for designating the humidity conditions of an atmosphere. One of these is the actual vapour pressure of the atmosphere. This should not be confused with the saturation vapour pressure. Since the rates of diffusion and evapouration are influenced directly by the vapour pressure of the atmosphere this is usually the most satisfactory unit in which to express humidity values for physiological purposes.

The vapour pressure deficit is also a common, and for many purposes a valuable unit for the expression of humidity values. The vapour pressure deficit is the difference between the actual vapour pressure of the atmosphere and the vapour pressure of a saturated atmosphere at the same temperature. As an example, consider an atmosphere having a vapour pressure of 10.51 mm. Hg at 250 C. The saturation vapour pressure of this atmosphere would be 23.76 mm. Hg. Hence its vapour pressure deficit is 13.25 mm. Hg (23.76 10.51). If an evapourating surface and the atmosphere are both at the same temperature the vapour pressure deficit indicates directly the steepness of the vapour pressure gradient between the atmosphere and the evapourating surface.

The term vapour pressure gradient as applied to water-vapour has the same significance as the term diffusion pressure gradient which has been frequently used in previous discussions. In fact the vapour pressure gradient of water vapour is its diffusion pressure gradient. As long as the temperature of the evapourating surface does not differ materially from that of an atmosphere the rate of evapouration from a saturated surface into that atmosphere will show a close proportionality to its vapour pressure deficit.

However, the temperature of an evapourating surface is seldom exactly the same as that of the atmosphere and the deviation in temperature between the two is often very considerable. There are therefore many situations in which rates of evapouration do not show a close proportionality with the vapour pressure deficit of the surrounding atmosphere. The most generally employed of all units in expressing humidity values is the relative humidity, which is the percentage saturation of an atmosphere at a given temperature. For example, a saturated atmosphere at 30 C. will have a vapour pressure of 31.82 mm. Hg.

The relative humidity of this atmosphere would be 100 per cent. If only half the amount of vapour is present that would be present at saturation at this temperature (*i.e.* a vapour pressure of 15.91 mm. Hg) then the relative humidity would be 50 per cent etc. The relative humidity depends both on the concentration of water-vapour in the atmosphere and upon the temperature. Any increase in temperature with no accompanying change in the amount of water-vapour present always results in a decrease in relative humidity, because an increase in temperature means that a higher vapour pressure will be required before the atmosphere is saturated.

Expression of humidity values in terms of relative humidity, although a common practice, is unsatisfactory for physiological purposes because the same relative humidity, 50 per cent for example, may refer to widely different vapour pressures and vapour pressure deficits. A relative humidity of 50 per cent at 10 C. is equivalent to a vapour pressure deficit of 4.60 mm. Hg, while a relative humidity of 50 per cent at 500 C. is equivalent to a vapour pressure deficit of 46.26 mm. Hg. In other words,with the same relative humidity, evapouration from moist surfaces exposed tothe air will be many times as rapid at 5Q0 C. as at 100 C. Only when all of the relative humidity values are recorded at the same temperature are they an expression of the relative differences in the vapour pressure of an atmosphere.

In general the greater the vapour pressure of an atmosphere, other factors remaining unchanged, the slower the rate of transpiration. The rate of diffusion of water-vapour out of a leaf depends upon the difference between the vapour pressure in the intercellular spaces and the vapour pressure of the outside atmosphere, since the vapour pressure is a measure of the diffusion pressure of the water-vapour. Let us suppose that the vapour pressure of the intercellular spaces is 31.82 mm. Hg which is the value for a saturated atmosphere at 30 C. Such a vapour pressure is frequently attained in the internal air spaces of leaves.

Let us further assume that at the same time the vapour pressure of the atmosphere is only half as great (15.91 mm. Hg). In very quiet air the vapour pressure in the neighbourhood of transpiring leaf surfaces may be greater than that in the atmosphere in general but in this discussion it is assumed that there is sufficient air movement to prevent any appreciable accumulation of water-vapour in the vicinity of the leaves. Under the conditions as stated diffusion of water-vapour would occur through the stomates at a relatively rapid rate. If the vapour pressure of the atmosphere were lowered below this value the rate of diffusion of water-vapour out of the leaf would be increased; conversely increase in the vapour pressure of the atmosphere would result in a decrease in the diffusion rate of water-vapour out of the leaf.

Similarly, an increase in the vapour pressure of the intercellular spaces would result in an increase in the rate of transpiration while a decrease in the vapour pressure of the intercellular spaces relative to that of the atmosphere would have the converse effect. On the rare occasions when the vapour pressures of the atmosphere and of the intercellular spaces are equal, no transpiration will occur, even if the stomates are open.

The final statement in the preceding paragraph, should not, however,be interpreted to mean that transpiration can never occur into a saturated atmosphere. If the temperature of the leaf is higher than that of the surrounding atmosphere this will, if the intercellular spaces be saturated with water-vapour, result in~ a higher vapour pressure in them than in the atmosphere, even when the latter is also saturated with water-vapour. Under such conditions outward diffusion of water-vapour occurs, resulting in a localized region of supersaturated atmosphere around the leaf in which condensation of water-vapour may take place.

CLASSICAL STUDIES ON PHLOEM TRANSPORT

In classic experiments on the translocation of organic solutes performed by the Italian anatomist Marcello Malpighi in 1686, the bark of a tree was removed in a ring around the trunk. In 1928, T. G. Mason and E. J. Maskell observed that this treatment, called girdling, has no immediate effect on transpiration, since water moves in the xylem, interior to the bark. However, sugar transport in the trunk is blocked at the site where the bark has been removed. Sugars accumulate above the girdle—that is, on the side toward the leaves—and are depleted below the treated region.

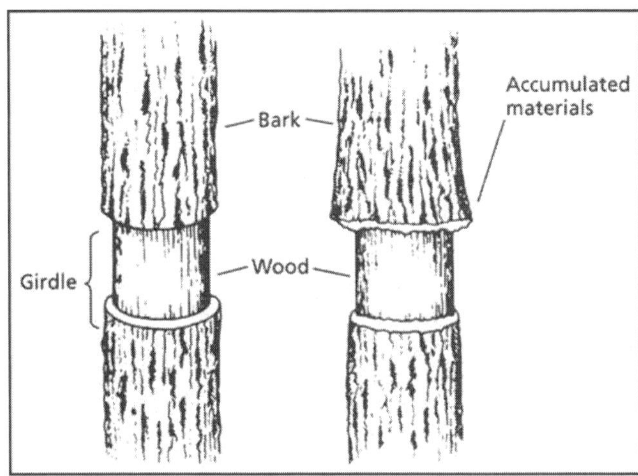

Fig. Tree Trunk Immediately after Girdling (left) and later (right).

Girdling is the removal of the bark of a tree in a ring around the trunk. At right, materials translocated from the leaves have accumulated in the region above the girdle and caused it to swell.

Eventually the bark below the girdle dies, while the bark above swells and remains healthy. Mason and Maskell concluded that sugar is transported in the bark of the tree and that the sieve elements are the cellular channels of sugar transport. The latter conclusion was based on an observed high correlation between leaf and bark sucrose contents and on the high sucrose concentration calculated to be present in the sieve elements.

More sophisticated experiments on phloem translocation became possible in the 1940s, when radioactive isotopes became available for scientific research. Labeled organic compounds can be introduced into the plant in various ways.

For example, carbon dioxide can be labeled with C or C and supplied to an intact mature leaf enclosed in a sealed chamber. These two isotopes of carbon, C and C, have very different half-lives (the half-life is the time required for the radioactivity to decrease by one-half). The half-life for C is 20 minutes; that for C is 5,600 years. For this reason, the isotopes are used in different types of experiments. Carbon-14 is useful for long term experiments analyzing the compounds in which the radioactivity is present.

Carbon-11 would disappear before such an analysis could be made. Carbon-11 is most useful when the researcher wishes to repeat measurements on the same plant. Since only a few hours are needed for the isotope in the plant to disappear, the radioactivity used in one experiment does not interfere with next.

The labeled carbon dioxide is incorporated into organic compounds via the CO_2-fixing reactions of photosynthesis and eventually into transport sugars. In another type of experiment, researchers bypass the photosynthetic reactions by supplying labeled transport sugar, usually sucrose, directly to the leaf.

To identify the pathway of sugar translocation, we must determine the tissue or cellular location of the isotope, usually by means of a technique called autoradiography. In tissue autoradiography, tissue of labeled plants is rapidly frozen, freeze-dried, embedded in paraffin or resin, and sliced into thin sections. The sections are then coated with a film of photographic emulsion.

During a period of exposure, radiation from the radioactive compound fed to the plant exposes the film, and when the film is developed, silver grains appear wherever the label was present in the tissue. A comparison of the tissue section and the pattern of silver grains shows the location of the label in the tissue. In terms of the transport pathway for sugars, the label initially appears in the sieve elements of the phloem, confirming the results of the earlier experiments.

CO_2 was supplied to a source leaf of morning glory (*Ipomea nil*). C was incorporated into sugars synthesized in the photosynthetic process, which were then transported to other parts of the plant. The location of the label is revealed in the tissue cross sections by the presence of dark grains on the film. (A) shows a low magnification of the cross section of the stem (50×), revealing dark spots resulting from the silver grains in the film, shown in higher magnification (325×) in (B). The label is confined almost entirely to the sieve elements of the phloem.

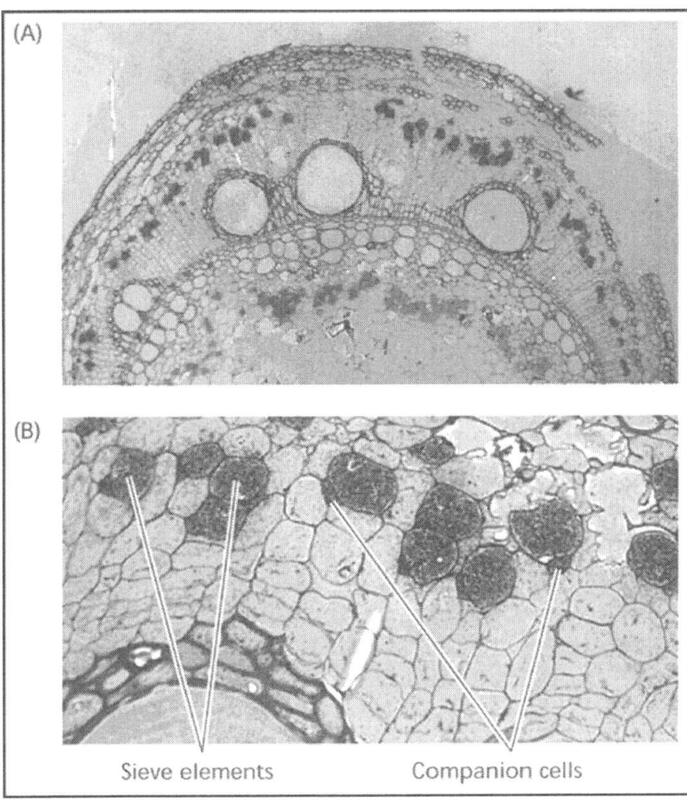

Fig. Photosynthate from the leaves Appears in the Sieve Elements of phloem in the stem.

NITROGEN TRANSPORT IN SOYBEAN

One of the most widely studied species in terms of nitrogen transport is the economically important soybean, a legume that has nitrogen-fixing root nodules. Research has shown that 12 to 48% of the nitrogen needs of the developing seeds of this species are supplied from compounds newly synthesized in the leaves, and 35 to 52% from compounds synthesized in the roots and transferred from xylem to phloem. The remaining nitrogen probably comes from the breakdown of specialized storage proteins in the soybean leaf.

In a recent study, non-nodulated soybean (*Glycine max* (L.) Merr.) plants were cultivated hydroponically under 5 m*M* $NaNO_3$ (ample N supply)) or 0.5 m*M* $NaNO_3$ (deficient N conditions). N or N labeled nitrate was fed to the cut end of the stems, and the accumulation of nitrate-derived N in the pods, nodes and stems was compared.

Real-time images of N distribution in stems, petioles and pods were obtained using a Positron Emitting Tracer Imaging System for a period of 40 min.

The results indicated that the radioactivity in the pods of N-deficient plants was about 10 times higher than that of plants that had sufficient N. The fact that the N translocation into the pods was much faster in N-deficient plants may be due to the strong sink activity of the pods in N-deficient plants.

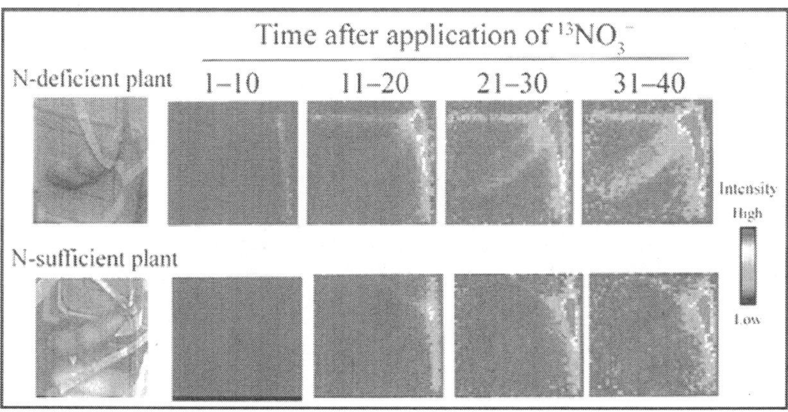

Fig. Images Detected by Positron Emitting Tracer Imaging System around the pod Position of the N-deficient and N-sufficient plants.

Each image was the average intensity during 10 min. Note that in the 21-30 and 31-40 min. images of the N-deficient pod, the pod image becomes noticeable as radioactivity accumulates in the pod. No comparable image can be seen in the plant with ample N supply.

SAMPLING PHLOEM SAP

Since phloem tissue, like xylem tissue, consists of a mixture of conducting and nonconducting cells, identifying the material translocated in the sieve elements is challenging. Methods based on labeling with radioactive isotopes and direct extraction of the tissue do not distinguish between compounds in transit and those outside the translocation pathway.

The tendency of some species to exude phloem sap from wounds that sever sieve elements has been exploited to obtain relatively pure samples of the translocated material.

The driving force for exudation is the high positive pressure in the sieve elements. However, most species do not exude detectable amounts of phloem sap, because of the sealing mechanisms closing wounds in the phloem.

However, exudation rates can be increased by treating the cut surface with the chelating agent EDTA (ethylenediaminetetraacetic acid). Callose synthase—the enzyme that synthesizes the sealing compound, callose—requires the presence of calcium, and since chelating agents bind divalent cations such as calcium very effectively, they lower the calcium concentration available for callose synthase.

The major disadvantage of the wound exudation method is that the fluid collected may not represent the true composition of the translocated material. Contaminants can originate from damaged parenchyma cells or even from the sieve elements themselves, and the sap is generally diluted by water influx from the xylem.

Furthermore, the abrupt lowering of turgor pressure in the sieve elements, causes a decrease in the water potential of these cells. As a result, water from the surroundings enters the sieve elements along a water potential gradient, causing the phloem sap to become diluted.

The ideal way to collect phloem sap would be to tap into a single sieve element by using a tiny syringe. Fortunately, nature has provided just such a probe: the aphid stylet. Aphids are small insects that feed by inserting their mouthparts, consisting of four tubular stylets, into a single sieve element of a leaf or stem.

The high turgor pressure in the sieve element forces the cell contents through the insect's food canal and into its gut, where amino acids are selectively removed and some sugars are metabolized. The excess sap, still rich in carbohydrates, is then excreted as "honeydew" (the droplet seen exuding from the body of the aphid in, which can be collected for chemical analysis.

Sap can be collected from aphid stylets cut from the body of the insect, usually with a laser, after the aphid has been anesthetized with CO_2. Exudate from severed stylets provides a more accurate picture of the substances present in the sieve elements than does honeydew (aphid excreta), the composition of which has been altered by the insect.

Experimental techniques using aphids have substantial advantages in the study of phloem physiology. Of particular importance is the fact that aphids tap a single sieve element, so there is no problem of contamination from other cell types. Exudation from severed stylets can continue for hours, suggesting that the aphid prevents normal sealing mechanisms from operating or that the drop in pressure that occurs when the sieve element is penetrated is not sufficient to trigger sealing mechanisms.

(A) An aphid (*Longistigma caryae*) feeding on a branch of linden (*Tilia americana*). The high turgor pressure in the sieve element forces the cell contents through the food canal of the aphid and into its gut. The honeydew excreted by the aphid consists of sieve element sap from which selected solutes, especially amino acids, have been removed in the gut of the insect. The high turgor pressure in the sieve element forces the cell contents through the food canal of the aphid and into its gut. The insect in the photo is releasing a droplet of honeydew, which it does about once every 30 minutes. (B)

Transverse section through the bark of linden, showing the tips of aphid stylets that had been exuding before they were sectioned. The tips of the stylets are inside a single sieve element; the sheath of saliva secreted by the aphid ends outside the cell.

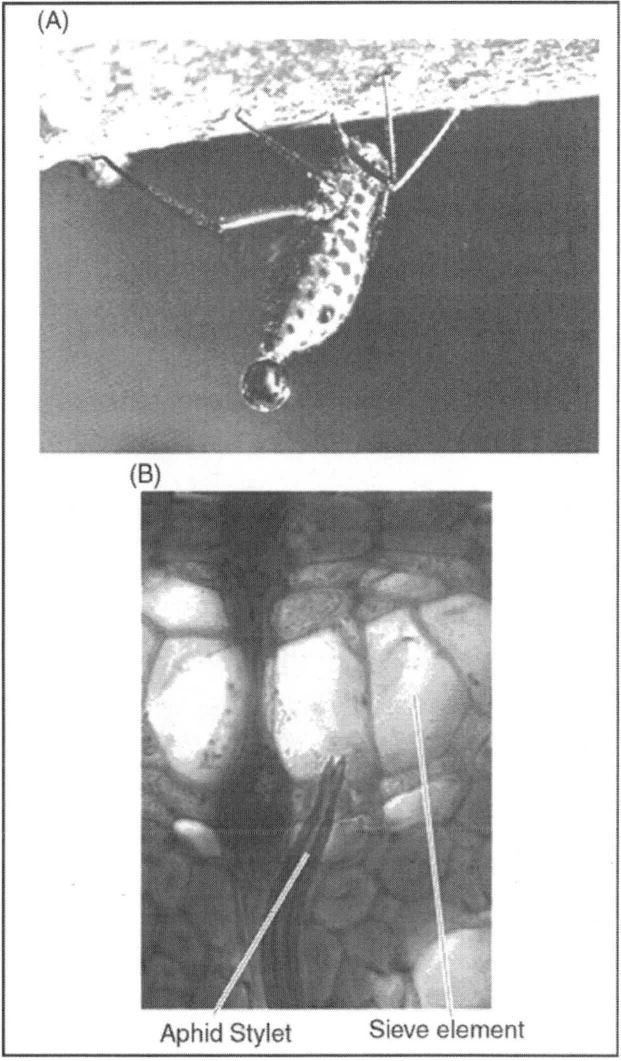

Fig. Using Aphids to tap a Single Sieve Element and collect Phloem Sap.

The disadvantages of using aphids include difficulties in placing the insects at a desired location and severing the stylets without disrupting them. In addition, aphids may induce reactions in the host plant by secreting saliva into the plant tissues.

However, the magnitude of these reactions over the period of time necessary to collect sap for analysis is probably negligible.

MONITORING TRAFFIC ON THE SUGAR FREEWAY

The speed of sugar molecules along the sieve elements of the phloem is known as the mass transfer rate. Measurements of mass transfer rates were among the earliest quantitative determinations made in phloem physiology. The early studies measured either gain in weight by developing fruits or storage organs (sinks) or loss of weight by mature leaves (sources) over a known time interval; the sink experiments are generally the more accurate.

For example, in 1926 Mason and Lewin determined weights of individual tubers of the greater yam (*Dioscorea alata*) over a growing season. The increase in dry weight was divided by the time interval to obtain an average mass transfer rate. This rate, divided by the average sieve element area per stem, yields a specific mass transfer rate of 5.7 g h cm of sieve elements. Several similar studies obtained mass transfer rates of 0.2 to 4.8 g h cm of phloem. If sieve elements are assumed to occupy approximately one-third of the phloem cross-sectional area (a reasonable estimate), these results yield specific mass transfer rates of 0.6 to 14.5 g h cm for sieve elements.

A more recent technique for determining mass transfer rates by use of radioactive tracers gives rates in excellent agreement with the earlier measurements. In this method the plant is usually pruned to a single mature source leaf and a single immature sink leaf. Carbon dioxide labeled with C is supplied to the source leaf in such a way that the rate of supply is equal to the rate of fixation.

The rate of arrival of label in the sink tissue is monitored for the duration of the experiment. The supplied radioactive CO_2 contains a predetermined amount of radioactivity per unit weight of carbon (specific activity). After a certain period of time, the sugars in transit from the source to the sink reach the same specific activity as the supplied CO_2. When this state, called *isotopic saturation,* has been reached, the mass transfer rate can be calculated from the rate of arrival of label in the sink tissue, as follows:

Arrival rate in sin k	Specific activity of sup plied label	Mass transfer rate

$$\frac{\text{Bq}}{\text{s}} \div \frac{\text{Bq}}{\mu\text{q of carbon}} = \frac{\mu\text{g of carbon}}{\text{s}}$$

where Bq stands for becquerel, a unit of radioactivity, and s stands for seconds. (A becquerel or (Bq) is equivalent to 1 disintegration per second. Another unit of radioactivity that has been used quite often is the curie or (Ci), which is equivalent to 3.7 x 10 disintegrations per second.) The rate determined in this way, corrected for the cross-sectional area of sieve elements in the transit pathway, provides an estimate of specific mass transfer. For sugar beet, which transports sucrose, the value obtained is equivalent to a rate of 4.8 g h cm of sieve elements, in excellent agreement with the values determined in earlier investigations.

EVIDENCE FOR APOPLASTIC LOADING OF SIEVE ELEMENTS

Early research on phloem loading focused on the apoplastic pathway. Many studies devoted to testing these predictions have provided solid evidence for apoplastic loading in several species. Some of the more important evidence is:

- Transport sugars are found in the apoplast. Sucrose is the predominant apoplastic sugar in species that transport mainly sucrose in the phloem, for example, sugar beet (*Beta vulgaris*) and broad bean (*Vicia faba*). Treatments and events that alter the rate of translocation from the source leaf also change the flux of sucrose through the apoplast or the apoplastic sucrose level.
- Sucrose supplied exogenously to a source leaf of sugar beet accumulates in the sieve elements and companion cells of the minor veins, as does sucrose derived from photosynthetic CO_2 fixation. Similar observations have been made in broad bean, pea (*Pisum sativum*), castor bean (*Ricinus communis*), and other species.
- PCMBS (*p*-chloromercuribenzenesulfonic acid) is a reagent that inhibits the transport of sucrose across plasma membranes but does not enter the symplast. PCMBS inhibits the uptake of sucrose from the apoplast when the sugar is supplied exogenously to sugar beet. Of greater importance, PCMBS also inhibits the export of sucrose synthesized from CO_2 in the mesophyll, implying that short-distance transport in sugar beet normally includes an apoplastic step. Assimilate loading is *also* inhibited by PCMBS in broad bean, pea, and other species.

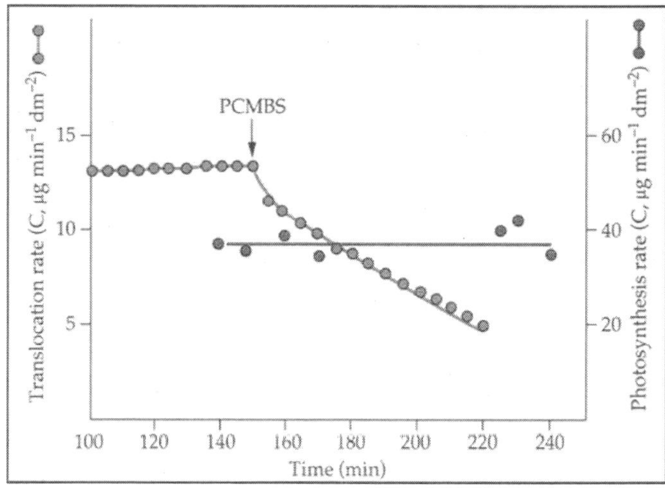

Fig. Effects of PCMBS on Phloem Transport from a Sugar Beet Leaf.

Export pf labeled photosynthate from a sugar beet leaf decreases upon addition of PCMBS. Photosynthesis was not affected, indicating that PCMBS did not alter the photosynthetic properties of the leaf.

- Experiments with transgenic plants further support the interpretation that sucrose moves through the apoplast. (Transgenic organisms carry cloned genes integrated into their DNA by a variety of recombinant-DNA techniques.) Transgenic tomato plants that have a sucrose-cleaving enzyme (invertase) in their apoplast show a very slow growth rate and fail to mobilize the starch in their source leaves during a prolonged dark period. Apparently, hydrolyzing sucrose in the apoplast inhibits phloem loading in source leaves of this species. This result is expected only if the phloem-loading pathway includes an apoplastic step.

SOME SUBSTANCES ENTER THE PHLOEM BY DIFFUSION

Many substances, such as organic acids and plant hormones, are found in the phloem sap at lower concentrations than carbohydrates. These substances are probably not actively loaded into the sieve element–companion cell complex but enter the sieve elements via other pathways and mechanisms. They may be taken up directly by diffusion across the phospholipid bilayer of the plasma membrane of the sieve element–companion cell complex or by a passive transporter in the plasma membrane of those cells, or they may diffuse into the sieve elements via the symplast.

Once in the sieve elements, these substances are swept along in the translocation stream by bulk flow, the motive force being generated by the active loading of only certain sugars or amino acids. Many substances not normally found in plants, such as herbicides and fungicides, can be transported in the phloem because of their ability to diffuse through membranes at an intermediate rate.

In other words, they diffuse through membranes rapidly enough to allow considerable accumulation in the sieve elements, but slowly enough that they are not lost from the sieve elements completely before reaching a sink tissue.

In addition, compounds that are weakly acidic tend to be "trapped" within the sieve elements, becoming negatively charged in the basic environment (low H concentration) of the sieve elements and thus less likely to diffuse out of the cells across the hydrophobic membrane. Substances that are not transported in the phloem (such as calcium) may not enter the sieve elements at all.

A recent study has characterized the *Arabidopsis* gene *AUX1* (auxin influx carrier) that encodes an auxin permease that appears to functin as an auxin carrier. *AUX1* has been localized in protophloem cells in the root. Gene expression studies have shown that *AUX1* facilitates IAA loading into the leaf vascular transport system, and IAA unloading in the primary root apex and developing lateral root primordium.

LOCALIZATION OF THE SUCROSE–H SYMPORTER IN THE PHLOEM OF APOPLASTIC LOADERS

SUC2 is one of the several sucrose transporters that have been localized in the phloem. As shown in Figure, SUC2 is found in the companion cells.

Mutant *Arabidopsis* plants containing transferred DNA insertions in the gene encoding SUC2 have been recently identified by reverse genetics. In the homozygous state, these mutations resulted in stunted growth, retarded development, and sterility. The source leaves of mutant plants contained a great excess of starch, and radiolabeled sugar failed to be transported efficiently to roots and inflorescences. Sucrose transporters are postulated to load photosynthetically produced sucrose into the sieve elements.

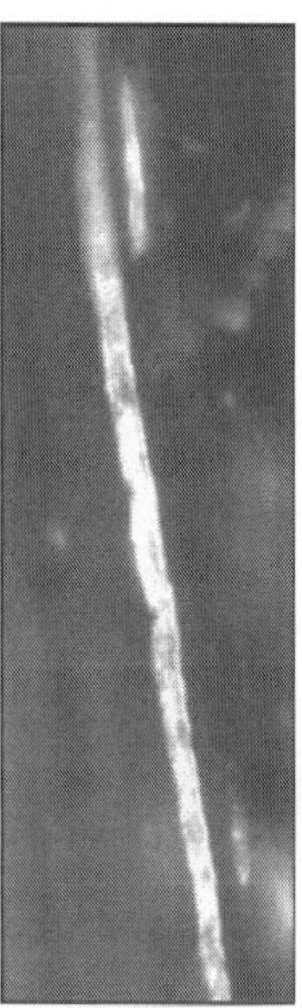

Fig. This Micrograph shows a single companion cell from Broad-leaved Plantain (*Plantago major*) Stained with two Fluorescent dyes.

One of the dyes (green) is (indirectly) linked to an antibody that is specific for the PmSUC2 sucrose–H symporter. The second dye (blue) binds to DNA. Since the two dyes are found on a single phloem cell, which is always adjacent to a sieve element, the sucrose symporter is concluded to be located in the companion cell membrane in this species. Interestingly, expression of the SUC2 transporter begins in the tip and proceeds to the base in developing leaves during a sink-to-source transition, the same pattern shown by photosynthate export capacity.

TEMPERATURE EFFECTS ON TRANSPIRATION

THERMAL RELATIONS OF LEAVES

While leaf temperatures often do not deviate greatly from surrounding atmospheric temperatures the discrepancy between the two is frequently sufficiently great to make it necessary to take it into account in careful experimental work. Leaves exposed to direct solar or artificial radiation usually have temperatures of from 2 to 100 C. (sometimes even more) in excess of that of the atmosphere. Under other conditions, to be described later, leaf temperatures may be less than that of the atmosphere.

Leaf temperatures are generally measured by means of thermocouples. A thermocouple is made by twisting together the ends of two fine wires of dissimilar.

Theoretically the temperature of a leaf may be regarded as conditioned by four different influences:

- Thermal absorption,
- Thermal emission,
- Internal endothermic (energy-storing) processes, such as photosynthesis and transpiration, and
- Internal exothermic (energy-releasing) processes such as respiration.

The influence of this last factor upon leaf temperatures is practically always negligible and will be disregarded. Similarly the quantity of energy transformed in photosynthesis is relatively so small that it need not be considered in evaluating the factors determining the temperature of leaves. Thermal emission refers to the loss of heat from a leaf by the processes of conduction, convection, and radiation. Thermal absorption refers to the gain of energy by a leaf by these same physical processes. Heat transmission which is brought about by intermolecular contacts is known as (thermal) conduction. The greater the temperature difference between a leaf and its environment the more rapidly conduction of heat will occur from the leaf to the gases of the atmosphere, if its temperature is the higher, or in the opposite direction if the temperature of the atmosphere is the higher.

Whenever loss of heat is occurring by conduction from a leaf to adjacent gas molecules of the atmosphere convection currents are set up in the atmosphere in the vicinity of the leaf. Cooler gas will displace the gas in the vicinity of the leaf surfaces which has become warmed as a result of thermal conduction from the leaf. This accelerates the rate at which conduction can occur from the leaf.

Radiation is the transfer of radiant energy across space. The best known types of radiant energy are light, infrared radiations (heat waves) and ultraviolet radiations. Radiant energy is commonly pictured as being propagated across space in the form of undulatory waves. Radiation occurs from the molecules of one body to those of another only if the radiating body is at a higher temperature than the receiving body. Thus light infrared and ultraviolet radiation are transferred from the sun to the earth metals, for example copper and constant an (an alloy of copper and nickel). A difference of electrical potential is set up between two wires thus brought into intimate contact. The magnitude of this electrical potential is very nearly directly proportional to the temperature of the thermocouple. In actual practice two junctions are generally used, one being inserted into the leaf blade, the other being kept at a standard temperature, usually 0 C. The difference in potential between the two thermocouples is generally measured with a potentiometer.

TEMPERATURE EFFECTS ON TRANSPIRATION

In like manner a warm stove loses heat by invisible infrared radiation to its environment. Radiation occurring from leaves is also in the infrared range of wave lengths. The temperature of a leaf exposed to direct sunlight or strong artificial illumination almost invariably exceeds that of the atmosphere. Thermal absorption in this case direct absorption of solar radiations proceeds under such conditions at a rapid rate. A portion of the absorbed radiant energy is dissipated (usually) by transpiration, and a portion is lost from the leaf by thermal emission.

As already shown transpiration alone is never adequate to dispose of the energy absorbed if the leaves are exposed to strong insolation. Thermal emission under these conditions will involve loss of heat both by radiation and by conduction. In general the greater the excess of leaf temperature over that of the atmosphere the larger the proportion of heat loss by thermal emission as contrasted with transpiration. On a cloudy day the temperature of leaves seldom deviates very greatly from that of the enveloping atmosphere. Heat exchanges between a leaf and its environment under such conditions probably occur principally by conduction. Similarly at night leaf temperatures usually do not deviate greatly from those of the surrounding atmosphere.

Leaves sometimes have a lower temperature than the surrounding atmosphere. This is often true, for example, of leaves which are transpiring fairly rapidly but which are not exposed to direct sunlight. Such leaves are often cooler by several degrees centigrade than the atmosphere. It is also possible that leaves sometimes lose heat energy by direct radiation to their environment rapidly enough to result in a lowering of their temperature below that of the atmosphere. This is especially likely to occur on clear nights when the vapour pressure of the atmosphere is low. These conditions favour direct radiation from the leaves to the sky, *i.e.* to relatively cold gases especially carbon dioxide and water-vapour. The temperature of leaves usually fluctuates from minute to minute especially during daylight hours. Minor fluctuations are due largely to shifts in wind velocity. The thermal emissivity of leaves exposed to direct sunlight is invariably increased by an increase in wind velocity. When the sunlight is intermittent, frequent changes in leaf temperatures also occur. Each time the sun is obscured by a cloud a sudden drop in leaf temperature usually takes place.

Contrariwise, when the sun emerges from behind a cloud there is usually a distinct and abrupt rise in leaf temperature. The factors controlling the internal temperatures of the other organs of plants are in general similar to those influencing leaf temperatures. The temperatures of fleshy leaves, fruits, tree trunks, succulent stems such as those of cacti, etc. under the influence of direct insolation may often greatly exceed those of the surrounding atmosphere. The side of an apple fruit exposed to direct sunlight, for example, may have a temperature of from 12 to 25 C. higher than the air temper-ature.

THE INFLUENCE OF TEMPERATURE UPON TRANSPIRATION RATES

The effectsof temperature upon the rate of stomatal transpiration can be most clearly analysed in terms of its effect upon the difference in vapour pressures between the intercellular spaces and the outside atmosphere. In this part of the discussion it is again assumed that the air movement is sufficient to prevent any appreciable accumulation of water-vapour in the vicinity of the leaf surfaces.

Suppose that the temperature of both the leaf and the surrounding atmosphere increases from 20 C. to 300 C. Unless the leaf is markedly deficient in water this will result in an increase in the vapour pressure of the intercellular spaces from approximately 17.54 mm. Hg to approximately 31.82 mm. Hg, these being the values for a saturated atmosphere at 200 C. and 300 C., respectively. The atmosphere of the leaf intercellular spaces is in direct contact with the relatively extensive evapourating surface of the mesophyll cell walls, hence the vapour pressure in the intercellular spaces tends to remain in equilibrium with the water in the mesophyll cells. In order to simplify this part of the discussion, it will be assumed that the atmosphere of the leaf intercellular spaces maintains essentially a saturation vapour pressure for the prevailing leaf temperature. Under certain conditions, as shown later in this chapter, maintenance of even an approximately saturated atmosphere throughout the intercellular spaces probably does not occur, but for the time being this possibility will be disregarded.

In the surrounding atmosphere, however, vapour pressure conditions are very different. On clear days, that is, on the very type of day upon which the highest rates of transpiration occur, there is frequently little change in the vapour pressure of the atmosphere over land surfaces during the course of a single day 2 Evapouration into the atmosphere is insufficient to permit a rapid building up of the vapour pressure towards the value for a saturated atmosphere as the temperature of the air increases during the day. This statement should not be misinterpreted to read that the vapour pressure of the atmosphere is invariable. The magnitude of the vapour pressure of the atmosphere varies greatly from day to day and from season to season, depending upon the prevailing climatic conditions.

On cloudy or rainy days, the atmospheric vapour pressure is generally greater than on clear days during the same season; in the summer months it is generally greater than in the winter months, etc. Nevertheless the statement made above that on clear, bright days there is often little change in the vapour pressure of the atmosphere is essentially correct that the process of transpiration itself, occurring on a grand scale from a vegetation-covered area of the earths surface would be sufficient to increase the vapour pressure of the lower layers of the atmosphere during the daylight hours. The atmosphere is so vast, however, in relation to the amount of water-vapour lost by plants, that over short periods, transpiration has only a slight effect on its vapour pressure except in deep ravines or other local habitats in which movement of the air is restricted.

Increase in the temperature of the atmosphere does result in an increase in the speed of the water-vapour molecules present. If the volume of the atmosphere remained constant this would result in a small increase in its vapour pressure. But even this effect is never fully realised in the atmosphere because an increase in temperature also results in an expansion of the atmosphere entirely or largely offsetting its influence in increasing vapour pressure.

If we assume for the purpose of our specific example that the vapour pressure of the atmosphere at 200 C. was half that of a saturated atmosphere at that temperature S.77 mm. Hgthen the excess vapour pressure of the leaf over that of the atmosphere at 30 C. was 8.77 mm. Hg (17.54 8.77). At 300 C., however, the vapour pressure of the intercellular spaces would have increased to about 31.82 mm.

Hg while the increase in the vapour pressure of surrounding atmosphere would usually be so slight that it can be disregarded. The excess vapour pressure of the intercellular spaces over the atmosphere is now 23.05 mm. Hg (31.82 8.77) which will result in diffusion of water-vapour out of the leaf at a rate nearly three times as fast as at 200 C. The effect of a rise in temperature therefore is principally an increase in the steepness of the diffusion gradient (vapour pressure gradient) of water-vapour through the stomates, and hence an increase in the rate of transpiration.

So far we have considered only examples in which the temperature of a leaf and the surrounding atmospheres are the same. We have already noted however, that the temperature of leaves exposed to direct sunlight is almost always higher than that of the atmosphere. If the temperature of a leaf is increased above that of the surrounding atmosphere by the absorption of solar radiation, the usual effect is an increase in the magnitude of the excess vapour pressure of the intercellular spaces over that of the outside atmosphere.

Let us continue the example which has been presented earlier. At 30 C. under the conditions as stated the vapour pressure difference between the intercellular spaces and the atmosphere was about 23.05 mm. Hg. Suppose, however, that due to the absorption of radiant energy, the temperature of the leaf is 350 C.while that of the atmosphere remains at 300 C. Water would evaporate from the walls of the mesophyll cells until the vapour pressure in the inter cellular spaces approximates that of a saturated atmosphere at 350 C. (42.18 mm. Hg).

The gradient between the intercellular spaces and the atmosphere is therefore increased to about 33.41 mm. Hg (42.18 8.77) and the rate of transpiration is correspondingly increased. The sudden increases in transpiration rate which are often observed when plants are shifted from shade to direct sunlight, as for example in potometer experiments, are undoubtedly due largely if not entirely to effects of this type. Similarly it has been observed that a sudden decrease may occur in the transpiration rate when a passing cloud obscures the sun, which is followed by just as sudden an increase when the sun again emerges from behind the cloud. For further analysis of the effects of leaf temperature upon transpiration rates.

WIND

The effect of wind upon the rate of transpiration is far from simple, and depends in part upon the other prevailing environmental conditions. Usually, however, increase in wind velocity, within limits, results in an increase in the rate of transpiration. This is usually explained by assuming that water-vapour often accumulates in the vicinity of transpiring leaves in a quiet atmosphere, especially if they are not exposed to direct sunlight. The result of such an accumulation of water-vapour is a decrease in the steepness of the vapour pressure gradient through the stomates and hence a decrease in the rate of transpiration.

If, however, the leaves are exposed to a wind, any accumulation of water-vapour molecules in the immediate vicinity of the leaf surfaces will be dispersed. The effective result will be an increase in the steepness of the vapour pressure gradient through the stomates, and a consequent increase in the rate of loss of water-vapour. Whenever a temperature differential exists between a leaf and the surrounding atmosphere, convection currents are set up in the gases in the vicinity of a leaf which may largely or entirely prevent any accumulation of water-vapour in the immediate vicinity of the leaf. The influence of wind in raising the transpiration rate of leaves is probably more effective, therefore, when they are subjected to such conditions that their temperature does not depart appreciably from that of the surrounding atmosphere.

The effect of wind in dispersing accumulated water-vapour in the vicinity of the leaf surfaces is probably of less importance in its influence upon transpiration rates than its effect in causing the swaying of branches and shoots and the bending, twisting, and fluttering of leaf blades. It has been shown experimentally that immobile leaves usually transpire less than similar leaves allowed to bend and move freely when both are exposed to wind of equal velocity. Such bending and contortion of leaves may increase the rate of water-vapour loss in part by compressing the intercellular spaces, thus forcing water-vapour and other gases out through the stomates.

INFLUENCE OF THE LIGHT FACTOR IN STOMATAL OPENING AND CLOSING

The most familiar of all stomatal reactions is their response to light. Unless other conditions, to be discussed later, are limiting, the stomates of most species open when exposed to light, and close upon the failure of illumination. Most commonly, therefore, the stomates are open in the daytime and closed at night, although there are many exceptions to this statement. The sensitivity of stomates to the light factor probably varies considerably according to species.

Within limits the stomates appear to respond quantit-atively to the amount of light absorbed. Stomatal opening apparently will occur in all wavelengths of the visible spectrum, although the influence of radiations in the red region appears to be weaker than the influence of other wavelengths of the visible spectrum. Upon the cessation of illumination the stomates usually begin to close. Generally this is a gradual process and, according to Stiil felt the greater the quantity of light which has been absorbed during the course of a day, the longer it takes for the completion of stomatal closure.

The work of Sayre Scarth and others indicates that the mechanism whereby light brings about stomatal opening and the mechanism whereby its absence causes stomatal closure are primarily osmotic, but are conditioned by changes in the H-ion concentration of the guard cells. Illumination of the guard cells has been found to result in a decrease in their H-ion concentration; failure of illumination in an increase. Thus Scarth found the pH of the guard cells of the Wandering Jew (Zebrina pendula) to range from 5.0 or less in the dark to between 6.o and 7.4 in the light.

The other cells of the leaf do not change appreciably in H-ion concentration upon the advent or failure of illumination, suggesting that the guard cells are less effectively buffered than the other leaf cells. Although several explanations have been offered to account for the effects of light and darkness on the pH of the guard cells none of them are strongly supported by experimental evidence and they will not be discussed. The guard cells apparently always contain starch, but the quantity present is not constant from one hour of the day to the next.

Sayre has shown that the starch content of the guard cells is at its maximum during the night, decreases rapidly during the daylight hours, and increases again towards evening.

The sugar content of the guard cells bears a reciprocal relationship to the starch content; when the latter is high, the sugar content is low, and vice versa. These reciprocal changes are apparently the result of reversible reactions in which the total amount of carbohydrate involved doesn't vary greatly. Conversion of starch to sugar and of sugar to starch results from the action of the complex of enzymes known as diastase. Increase in the pH of the guard cells such as occurs upon the incidence of light, appears to favour the hydrolytic (starch to sugar) action of diastase. On the contrary a decrease in their pH, such as occurs in the evening, favours the synthetic action of this enzyme whereby sugar is converted into starch.

Increase in the sugar concentration of the guard cells results in an increase in their osmotic pressure while a decrease in their sugar concentration has the opposite effect. That such changes in the osmotic pressure of the guard cells actually occur has been shown by many investigations. The diurnal changes in the osmotic pressures at incipient plasmolysis of the guard cells and epidermal cells. In general the osmotic pressure of the guard cells is usually relatively high during the daylight hours, and relatively low at night.

The osmotic pressure of the epidermal cells does not change appreciably during the course of the day and approximates that of the guard cells at night. Increase in the osmotic pressure of the guard cells in the morning results in an increase in their diffusion pressure deficit relative to that of the contiguous cells. Water therefore moves into the guard cells, increasing their turgor, which in turn leads to a widening of the stomata1 aperture. Similarly a decrease in the osmotic pressure of the guard cells results in a diminution of their turgor and a narrowing of the stomata1 aperture. However, certain facts suggest that the effects of light upon stomates cannot be interpreted entirely in terms of the osmotic mechanism just described.

One of these is the rapidity with which stomatal opening occurs. In some species at least the stomates open within less than a minute after exposure to light. It is difficult to visualize such a rapid action in terms of an enzymatic reaction, since usually such reactions occur at a relatively slow rate. Some investigators consider therefore that light exerts a direct effect upon the guard cells in addition to its indirect influence upon the magnitude of their osmotic pressure. Little if anything is known, however, concerning the mechanism of any such reaction.

As subsequent discussion will show development of an internal water deficit in plants is of frequent occurrence, especially on warm summer days. A shrinkage in the total volume of water in a plant results in general in a diminution in the volume of water in each individual cell, although all cells will not necessarily be affected equally. Such a decrease in the water content of the leaf cells, not sufficient to induce visible wilting, is often called incipient wilting.

Under such conditions the guard cells usually decrease in turgor as a result of osmotic movement of water into contiguous cells. Reduction in the turgor pressure of the guard cells as a result of the diminution of the volume of water in them will bring about a partial to complete closure of the stomates. There is also some evidence that diminution in the water content of the guard cells induces a decrease in the pH of their cell sap and a correlated conversion of sugar to starch. The resulting diminution in the osmotic pressure of the guard cells may lead to further loss of water from the into adjacent epidermal cells.

THE MAGNITUDE OF TRANSPIRATION

The Magnitude of Transpiration whether computed on the basis of a unit area of leaf surface, an entire plant or an acre of forest or crop plants, shows a great variation depending both on the plant species and environmental conditions to which they are exposed. The quantity of water transpired may be computed and compared for hourly daily, seasonal, or yearly periods. Because of the great fluctuations in transpiration rates with environmental conditions the citation of isolated values of transpiration rates is usually of very little physiological significance.

It is impossible to calculate accurately the transpiration rate of large plants such as trees from the known rate of transpiration of some of the leaves and an estimate of the area of all of the leaves on the tree. In the first place it is difficult to arrive at any accurate estimate of the aggregate leaf area of a large tree. In the second, there is a great variability in the transpiration rate among the different leaves. The rate of transpiration of the individual leaves varies depending upon their position, exposure, age, and internal physiological conditions.

For example, Huber found that low twigs of a redwood tree transpired six times as rapidly as similar twigs from the height of 12 metres. Neither can it be assumed, as mentioned in the discussion of the potometer method, that the transpiration rate of leaves of branches which may have been removed from a tree and placed with their cut ends in water will be the same as their rate of transpiration before the branches were detached.

Hence attempts to determine the transpiration rate for an entire tree from values determined for a few of the leaves on that tree can only result inapproximations.

The same comments in general apply, although perhaps not so emphatically, to calculations of the total transpiration of fields of cultivated crops and natural vegetation areas based upon transpiration as determined for a limited number of individual plants.

If transpiration is measured for one minute intervals rates much in excess of this may be observed, but such rates are maintained only for relatively short periods of time of necessity approximations, although, not of course, without their utility in the consideration of certain types of problems. The results of such calculations indicate that sufficient water may transpire from maize plants during the course of a season to cover the field in which they are growing to a depth of 15 inches.

Similarly Minckler has calculated that in a growing season water equivalent to 4.8 inches of rainfall may transpire from an acre of American elm trees and that an acre of red maples, growing in a very moist habitat, may lose water equivalent to 28.3 inches per acre. In evaluating all such data it must be remembered, however, that the magnitude of transpiration varies tremendously with the available soil water supply.

SIGNIFICANCE OF TRANSPIRATION

Opinions regarding the significance of transpiration have ranged all the way from those which would put it practically on a par with such processes as photosynthesis and respiration, to those which would relegate it to the category of a necessary evil. The principal roles which have been ascribed to the transpiration of plants can be summarized under the following three headings: Supposed Role in the Movement of Water.

It is often claimed or assumed that the movement of water through the plant requires the occurrence of transpiration from the leaves. That this concept is entirely erroneous will be clear from the discussion of the mechanism of the conduction of water through plants. Under conditions of high transpiration the movement of water through plants is, in general, more rapid than under conditions of low transpiration. The mechanism causing ascent of water through a plant operates in such a way that any decrease in the turgidity of the mesophyll cells favours a more rapid movement of water towards those cells. Hence a rapid transpiration rate, which invariably results in a considerable loss of turgidity by the mesophyll cells, usually speeds up the rate at which water ascends through the plant. However translocation of water to the extent that it is used in photosynthesis, growth or other metabolic processes continues even if the transpiration rate is negligible.

Supposed Role in the Absorption and Translocation of Mineral Salts. It has often been assumed that the more rapid the rate of transpiration, the greater the rate of absorption of mineral salts. This view implies that the dissolved mineral salts are swept into the plant along with the absorbed water, a postulation which ignores much evidence that the mechanisms operating in the absorption of water and the absorption of mineral salts are very different. The results of certain experiments do indicate that a somewhat larger quantity of mineral salts accumulates in plants under conditions favouring high transpiration than in similar plants growing under conditions of low transpiration although there is no actual proportionality between the volume of water absorbed and the quantity of mineral salts which pass into the plant.

Since even when plants grow under conditions favouring low transpiration rates they usually obtain an adequate quota of the various mineral salts, provided they are present in sufficient abundance in the soil, it is difficult to see that this effect is of any very great advantage to the plant. Realization is now fairly general that there is, at the most, only a slight correlation between the rate of transpiration and the rate of absorption of mineral salts from the soil.

It is also generally considered that transpiration plays an important part in the translocation of mineral salts from the root system to the top of the plant. That some upward translocation of mineral salts occurs in the transpiration stream is indisputable, but whether this is an important part of the total quantity transported is open to debate. There is considerable evidence that at least a part of the mineral salts which move through the plant are translocated in the phloem. The rate of transpiration can have little or no direct effect upon the movement of such salts. While high transpiration rates may mean a somewhat higher rate of transport of mineral salts through the plant than low ones it is doubtful if this is of great significance from the standpoint of the plant. Supposed Role in the Dissipation of Radiant Energy.

Leaves exposed to direct sunlight absorb large quantities of radiant energy which, unless dissipated in some other way, will be converted into heat energy and raise the temperature of the leaf. The possibilities of such an effect are indicated by the following approximate calculations: In direct noon-day summer sunlight the rate of receipt of solar energy is often 1.3 g.-cal. per square centimeter of leaf surface per minute and not uncommonly even greater. The proportion of this actually absorbed varies with the kind of leaf, but will be assumed to be only 50 per cent, which is a conservative estimate.

The incident radiant energy which is not absorbed by the leaf is all transmitted or reflected. Such a small proportion of the absorbed energy is used in photosynthesis that it can be neglected in a rough calculation. Hence about 0.65 g.-cal. of energy is absorbed per square centimeter of leaf surface per minute. If the mass of a square centimeter of a leaf is taken as 0.020 g. and its specific heat as 0.879g.–cal.2 the rise in temperature per minute would be 0.60.020 $\times$ 0.879 or about 370 C. Since the thermal death point of most plant protoplasm lies. These are the actual values as determined for a sunflower leaf by Brown and Escombe.

It should not require more than two minutes to heat the leaves of most plants to a lethal temperature. Actual measurements of leaf temperatures, however, show that they seldom even approach their thermal death points. Leaf temperatures usually do not exceed atmospheric temperatures by more than a few degrees centigrade.

Evidently some efficient energy-dissipating mechanism is at work which prevents the accumulation of heat energy in leaves. Since transpiration is an energy-consuming process it has often been assumed that it is in the evapouration of water from the leaves that most of the energy absorbed by them is dissipated. We might well inquire, therefore regarding the possible efficiency of transpiration as an energy-dissipating process. The evapouration of a gram of water at 20 C. requires 584 g.-cal. For the dissipation of 0.6 g.-cal. of heat, therefore, the evapouration of 0.0011 g.(0.65) of water per square centimeter of leaf surface per minute is required.

This is equivalent to 6.6 g. (0.0011 $\times$ 100 $\times$ 60) of water per square decimeter of leaf surface per hour, a rate of transpiration which is seldom if ever attained by plants for any sustained period of time. Evidently transpiration even when occurring at its maximum rate, can at the most account for a dissipation of only part of the radiant energy which is absorbed by leaves indirect sunlight.

The fact that transpiration is often inadequate in direct sunlight to account for the dissipation of all the absorbed radiant energy leads naturally to the question of whether it is at all essential for this process. Observations have shown that leaves in which the transpiration rate is greatly reduced, as for example those in which the stomates are plugged with vaseline, leaves in a wilted condition, or the leaves of plants in xeric habitats during dry seasons in which the occurrence of any appreciable amount of transpiration is precluded by the lack of soil water, seldom have temperatures which are any where near the thermal death point, even when exposed to direct sunlight.

Although transpiration often accounts for the dissipation of some or even most of the absorbed radiation, in so doing it apparently plays no essential role since absorbed radiation can be dissipated by purely physical means. As soon as the temperature of a leaf exceeds that of the surrounding atmosphere, it will lose heat to the atmosphere in the same way that any other object heated above its environmental temperature does, *i.e.* by one or more of the purely physical processes of conduction, radiation, and convection. The term thermalemissivity is frequently applied to this physical loss of heat energy by leaves and other objects. Actually, instead of benefiting plants, transpiration may often be detrimental. Under conditions of deficient soil water or of high transpiration rates even when the soil water supply is adequate, this process results in adiminution in the water content of a plant and in the turgidity of its cells. Prolonged drought conditions ultimately result in a severe desiccation withthe consequent death of all except the most drought resistant species.

When the diminution in water content is less severe, a train of other effects such as a decrease in the turgidity of the cells, stomatal closure and reduction in rate or cessation of photosynthesis are induced, all of which have the end result of checking the growth of the plant. It is probably true that lack of water is more often the limiting factor in plant growth than any other single factor. Furthermore deficiency of water is probably responsible for the death of more plants under natural or even cultural conditions than any other single cause.

The fundamental effects of transpiration upon the plant are not to besought in any hypothetical advantages of the process to the plant, but in its very real and readily ascertainable influences upon the water relations of plant cells and tissues, and through these its effects upon other plant processes. In spite of the fact that transpiration may be regarded in a sense as an incidental phenomenon, its indirect influence upon the metabolic processes of plants is a profound one. It is this fact, more than any other, which justifies intensive and critical studies of this process.

THE PROCESS OF EVAPOURATION

The fact that transpiration may be regarded as essentially a modified form of the process of evapouration makes desirable a fuller consideration of the dynamics of this process. When an open pan of water is exposed to the atmosphere, the level of the water in the pan slowly drops. Evidently water molecules are being slowly lost into the atmosphere. This is an example of the well known process of evapouration.

All of the molecules in a mass of water are not travelling with the same velocity, although for any given temperature the average speed of all of the molecules in the liquid mass is a constant. Some of the molecules in the liquid water attain sufficient momentum to entirely overcome the attractive forces holding them in the liquid and escape into the surrounding atmosphere in the form of vapour. The swiftest molecules present are most likely to be able to overcome the attraction of the other water molecules, therefore they are the most likely to be lost from the liquid.

During evapouration any body of water is thus slowly being depleted of its more rapidly moving molecules. The residual water becomes progressively richer in relatively sluggish molecules; in other words it becomes cooler. This cooling effect is more or less completely offset, however, by physical transfer of heat into the water from its surroundings as soon as its temperature drops below that of the environment. The proportion of high speed molecules in the molecular population of the pan of water is thus maintained at close to its original value and the rate of evapouration continues with very little diminution.

In the preceding paragraph we have focussed our attention only on thewater molecules which escape from the liquid. Water-vapour molecules are also returning to the liquid from the atmosphere during the process of evapouration. If a water-vapour molecule in the atmosphere above the evapourating surface, particularly one of the more sluggish ones, strikes the surface of the water at the proper angle and with not too great a velocity, it will be held there by the attractive forces exerted by the liquid water molecules, and again become part of the liquid water. High speed molecules, on the other hand are much less likely to be captured by the attractive forces exerted by the molecules at the surface of the body of liquid. They impinge upon the bounding film of the liquid with such velocity that unless they hit that surface a stright angles or nearly so, they usually glance off along a new pathway.

This picture of the kinetics of the evapouration process holds for any evapourating surface whether it be the exposed surface of a body of water a moist piece of cloth, paper, or porous clay, or the mesophyll cell walls of a leaf. Ordinarily the term evapouration is used to refer only to conditions in which the rate of escape of molecules exceeds their rate of return, and its use will be restricted to this sense in this discussion.

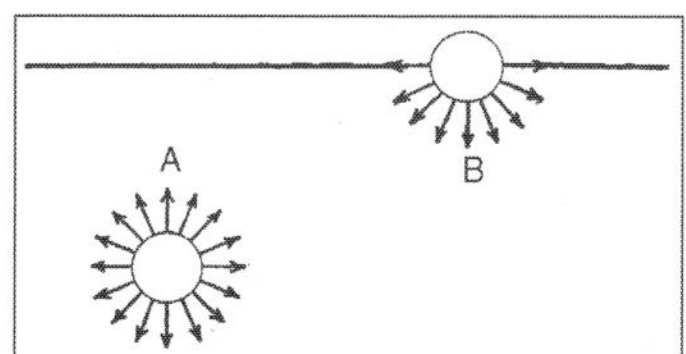

Fig. A Diagrammatic Representation of the Direction of the Forces of Cohesion Acting Upon a Water Molecule: (A) In the Body of the Liquid, (B) At its Surface.

If the air above the water surface is confined, as for example, when a dish of water is covered by a bell jar, the number of water-vapour molecules in the confined space will gradually increase due to evapouration from the surface of the water. Since their movement is a random one the water-vapour molecules will be continually colliding with walls of the container, each other the molecules of other gases present in the air, and surface of the liquid water. Some of those which strike the surface of the liquid will be held there by intermolecular attractive forces.

As the concentration of water-vapour molecules in the air increases, the number which plunge back into the water in any unit interval of time will increase. Eventually a dynamic equilibrium will be attained at which the number of molecules leaving the surface and the number returning to it in a unit time will be equal. At this point the airwill be saturated with water-vapour and evapouration will no longer be occurring. The water-vapour molecules exert a definite pressure against the walls ofthe container and the surface of the water. This is known as vapour pressure.

In the illustration given in the preceding paragraph, tihe vapour pressure of the water increases progressively until the saturation point is reached. The vapour pressure of water under the conditions of such a dynamic equilibriummay conveniently be termed the saturation vapour pressure. The vapour pressure of a liquid is usually expressed in terms of millimeters of mercury. The saturation vapour pressure of water at 2O C., for example, is I 7.54 mm. Hg.This means that at this temperature the water-vapour exerts a pressure equalto the pressure exerted by a column of mercury I 7.54 mm. high. The saturation vapour pressure of any liquid is independent of the area of the evapourating surface, but increases with increase in temperature.

Botanists are often interested in measuring the rate of evapouration undera given set of conditions in connection with studies of transpiration and of plant distribution. The difficulties involved in making this apparently simple measurement are not all apparent at first consideration. It would appear that the rate of evapouration could be measured by the rate at which water disappeared from an open pan exposed to the atmosphere. While such methods have been used they are subject to numerous errors and limitations.

The rate of evapouration for such a pan is controlled not only by environmental conditions, but also by the size, shape, and colour of the pan, and the depth of water below the rim. Other errors arise from the accumulation of rain water and the splashing of water due to wind or other causes. The many difficulties encountered in the use of open pans of water for measuring evapouration rates led to the devising of more suitable instruments for this purpose. Many measurements of evapouration rates have been made with instruments called atmometers.

These consist essentially of surfaces of porous clay moulded in the form of hollow cylinders or spheres.Water evapourates from such surfaces in essentially the same way that it evapourates from a free surface of water.

Atmometers are attached to a reservoir of water and are usually provided with mercury valves which prevent absorption of rain. The loss of water from such instruments can be determined either by measuring the decrease in volume of water or the loss in weight of the instrument.

Although atmometers are not subject to many of the errors inherent in the use of open-panevapourimeters numerous precautions must be observed in employing them. Details of atmometric technique are discussed by Livingston Measurement of Transpiration. Four general methods are in use for measuring transpiration, but only the first two of these as listed can be used for quantitative determinations of the rate of loss of water-vapour from plants which are rooted in the soil.

STRUCTURAL FEATURES OF PLANTS TRANSPIRATION

A number of diverse species of plants growing under identical environmental conditions usually differ very greatly in transpiration rates. Physiological factors, such as differences in stomatal behaviour, cell sap concentration, colloid content of the cells, etc. are undoubtedly partially responsible for the fact that all plants do not lose water-vapour at the same rate even under the same environmental conditions. Structural differences, particularly in the leaves, also account in part for unlike rates of transpiration of different species growing in the same environment.

Many of the supposed effectsof structural differences in plants upon their transpiration rates are, however based solely on inferences, and not upon experimentally determined facts. Many species which on the basis of their anatomy have been judged to have a low transpiration rate later were actually found to transpire very rapidly when environmental conditions were suitable. The following discussion of the anatomical features known or generally believed to influence the rate of transpiration will therefore be very brief.

THICKNESS OF THE CUTIN

It can be easily demonstrated that water vapourates more readily from uncutinized epidermal cell walls than from those which are coated with a layer of cutin. It is sometimes assumed, therefore that the thicker the layer of cutin on the outer walls of the epidermis of a leaf, the slower the rate of cuticular transpiration from that leaf. Actually,however, it seems obvious that any increased thickness in the cutin layer beyond that at which cuticular transpiration becomes negligible will have no appreciable influence on the rate of evapouration from the epidermal cell walls.

EPIDERINAL HAIRS

H airs are prominent features of one or both of the surfaces of the leaves of many species. Living hairs probably increase the rate of cuticular transpiration by increasing the exposed surface of the leaf. Many botanists have assumed that dead hairs reduce the rate of stomata1 transpiration, particularly under conditions of intense sunlight or strong winds. In the former case it has been supposed that the whitish hairs reflect such a large proportion of the incident light that the leaf temperature is not as high as it otherwise would be, and that the rate of transpiration is correspondingly lower. In the latter case it has been supposed that the hairs impose a mechanical barrier to the effect of wind on transpiration. However, Sayre found that removal of the hairs from the leaves of the mullein had little effect upon the rate of stomatal transpiration under ordinary conditions of light intensity and wind velocity, although such treatment did result in an increase in cuticular transpiration.

RATIO OF INTERNAL TO EXTERNAL SURFACE IN LEAVES

Although transpiration rates are most frequently expressed in terms of the exposed area of the leaf surface, most of the water-vapour lost from leaves evapourates from the walls of the mesophyll cells which bound the intercellular spaces. The area of this internal evapourating surface in proportion to the external surface of a leaf varies greatly not only in leaves of one species as compared to those of another, but in leaves of the same species if they have developed under different environmental conditions.

Computations of the ratio of internal exposed surface to external exposed surface of leaves of a number of species have yielded values ranging from 6.8 to 31.3. In general the greater the proportion of palisade to sponge tissue the greater this ratio.The thickness of the leaf will also obviously be a factor influencing this ratio.The ratio is greater in sun leaves than in shade leaves of the same species; for the lilac the values are 13.2 and 6.8, respectively.

It has generally been supposed that the transpiration rates of leaves in which a relatively large area of mesophyll cell walls is exposed to the intercellular spaces as compared with the exposed epidermal area of the leaf are greater than in leaves in which the contrary condition obtains. Theoretically such an effect could be exerted only if the area of exposed mesophyll cell walls exercised a controlling influence upon the steepness of the vapour pressure gradient through the stomates.

When transpiration is relatively high a steeper vapour pressure gradient may be maintained by leaves possessing relatively extensive internal evapourating surfaces than by those which do not. The higher transpiration rate of sun leaves as compared with shade leaves on the same plant, which has been observed in some species, may be partly explainable on this basis.

SUNKEN STOMATES

The stomates of many species of plants are sunken below the general level of the epidermis. Since the length of a diffusion gradient is one of the factors governing its steepness, diffusion through a small opening is slower if the gas must pass through a relatively long tube before reaching the orifice, than if the diffusion is through a shorter tube. In leaves in which the stomates are not sunken the diffusion column is relatively short, amounting only to the diameter of the guard cells as seen in cross section, while in sunken stomates the column of gas may be longer by several times.

The diffusion of gases through such stomates is undoubtedly slower than through stomates of the ordinary type. This retarding effect of sunken stomates upon the rate of transpiration as compared with the stomates of the non-sunken type has been estimated by Renner to range from 30 to 70 per cent.

STRUCTURE AND DISTRIBUTION OF THE ROOT SYSTEM

Under some environmental conditions an individual plant of a given species will develop a greater leaf surface in proportion to the extent of its root system thanunder others. The latter type of structural development will, other conditions being equal, favour the maintenance of higher transpiration rates thanthe former.

The distribution of a root system in the soil sometimes influences rates of water absorption and transpiration. In a habitat where deep-rooted and shallow-rooted species are growing side by side the former may transpire more rapidly during drought periods than the latter because their roots penetrate to a soil horizon which still contains available water, while roots of the latter are in a moisture deficient soil. The completeness with which the soil mass is interpenetrated by roots may also influence rates of absorption and transpiration.

Thus Miller found that sorghum plants have almost twice as many fibrous roots as maize plants. Rates of absorption and transpiration can be better maintained, especially when the soil is relatively dry, by plants with the sorghum type of root system than by plants with the maize type of root system.

THE DAILY PERIODICITY OF TRANSPIRATION

All plants exhibit a daily periodicity of transpiration rate which varies somewhat with the species and is greatly influenced by the environmental conditions to which the plant is exposed. We shall first consider the transpiration periodicity upon a standard day as defined in the preceding chapter, since most experiments upon daily variations in transpiration rate have been conducted upon plants growing under approx-imately such conditions.

It should not be assumed, however, that under natural conditions the daily periodicity of transpiration usually or normally follows the trend. The daily periodicity of transpiration under these roughly defined conditions does, however, afford a convenient reference standard with which to compare the many variations which are manifested by this phenomenon.

The transpiration rate during the hours of darkness is generally low, and in most species water loss during this period may be regarded as almost entirely cuticular. It is not justifiable to assume that absolutely no stomatal transpiration occurs at night in any species unless this is definitely known to be true. In some species of plants, as we have already seen, nightstomatal opening is a common occurrence, while in others it occurs under certain environmental conditions. Even in those species in which the stomates are normally closed during the hours of darkness, a few may remain open.

The transpiration rate shows a rapid rise during the early morning hours followed by a slower but consistent increase which culminates in a point of maximum transpiration which occurs sometime during the mid-day hours. Following this peak of transpiration, the rate decreases usually consistently although often with an accelerated rate in the late after noon, until the low and virtually steady night rate is attained at approximately the termination of the daylight period.

The daily periodicity in the rate of stomatal transpiration in most species of plants can be interpreted almost entirely in terms of two variable factors:

1. The diffusive capacity of the stomates, and
2. The vapour pressure gradient between the intercellular spaces of the leaf and the outside atmosphere. In analyzing the daily periodicity of transpiration it is the diffusive capacity of the stomates in the aggregate (*i.e.* all of the stomates on a plant) which must be considered. The aggregate diffusive capacity of the stomates on a plant on which all of the stomates were half open, for example, would be much greater than that for a similar plant on which half of the stomates were fully open and half fully closed.

With the advent of daylight the stomates open over a period of time which varies in length according to the species and with environmental conditions. Some open earlier or more rapidly than others. The initial rise in the rate of transpiration in the morning is due to the opening of the stomates, resulting in a gradual increase in their aggregate diffusive capacity. As each stomate opens a vapour pressure gradient is established through it between the atmosphere of the intercellular spaces and the outside atmosphere. During the hours of darkness the leaf cells regain most or all of the water which they have lost during the preceding day, and thus attain approximately their normal turgidity.

This facilitates the evapouration of water into the intercellular spaces, so that when the stomates open in the morning be internal atmosphere of the leaf is usually saturated with water-vapour, or practically so. The vapour pressure of the intercellular spaces is therefore at the maximum possible for that leaf temperature. This is, on clear days, almost always in excess of the vapour pressure of the atmosphere when the stomates open, and often considerably so. Hence outward diffusion of water-vapour through the stomates usually begins as soon as they are open. Once a stomate is well open, minor variations in the size of the stomatal aperture have little or no effect on the rate of water-vapour loss through it.The rate of transpiration continues to increase, however, for some time afterthe stomates, in the aggregate, have attained their maximum diffusive capacity. This is due to a gradual increase in the steepness of the vapour pressure gradient through the stomates. As the day progresses the temperature of both the atmosphere and the leaf increases; if the latter is in direct sunlight its temperature is invariably somewhat in excess of that of the atmosphere. On clear days, as previously described, the vapour pressure of the intercellular spaces usually increases relative to that of the atmosphere with a rise in temperature.

The steepness of the vapour pressure gradient between the internal atmosphere of the leaf and the external atmosphere therefore usually increases progressively during the earlier part of the day, and this is the important factor accounting for a rise in the transpiration rate once the stomates have attained approximately their maximum diffusive capacity. Almost from the moment when the stomates begin to open in the morning a train of events is set in operation in the plant which ultimately causes a reduction in the rate of transpiration.

During approximately the first half of the day, however, these factors are more than offset by the factors resulting in an increase in the rate of transpiration. In most plants, while transpiration is occurring rapidly, the rate of absorption of water does not keep pace with the rate at which water-vapour is lost from the leaves. This results in a reduction in the water content of the entire plant, and especially that of the leaves. In more extreme cases wilting results, but under standard day conditions the leaves seldom pass beyond the stage of incipient wilting which corresponds only to a partial loss of turgor by the leaf cells.

THE DAILY PERIODICITY OF TRANSPIRATION

Decrease in the water content of the leaf cells also results in a rise in their diffusion pressure deficit, and a correlated decrease in their vapour pressure. A fully turgid cell (*i.e.* one with a diffusion pressure deficit of zero) has an equilibrium vapour pressure equivalent to that of pure water regardless of its osmotic pressure.

At 20 C. this is 17.54 mm. Hg. If the diffusion pressure deficit of the mesophyll cells of a leaf at this temperature were raised to 100 atmos., a value which can be attained in only a very few species of plants, their vapour pressure would decrease by only about 1.27 mm. Hg. Since the vapour pressure of the cell walls, with which the vapour pressure of the intercellular spaces tends to be in equilibrium, in turn tends to be in equilibrium with that of the cell sap, it would appear that an increase in the diffusion pressure deficit of the mesophyll cells has only a very slight influence upon vapour pressure conditions in the intercellular spaces.

However, certain other factors must be taken into consideration in evaluating the effect of a diminished leaf water content upon the vapour pressure of the intercellular spaces. Decrease in the water content of the leaf cells involves a reduction in the quantity of water in the cell walls as well as in other parts of the cell. Water undoubtedly passes through cell walls principally through the intermicellar material. The principal effect of a decrease in the quantity of water in the walls is a shrinkage in the volume of the hydrophilic intermicellar material and a resulting contraction in diameter of the intermicellar capillaries.

This probably results in a decrease in the permeability of the walls to water. Hence, under such conditions, water often may not pass across the walls with sufficient rapidity that an equilibrium vapour pressure is maintained between the cell sap and the outer surface of the cell walls. When the water content of a leaf is low the vapour pressure of the cell wall surface in contact with the intercellular spaces may therefore be less than the vapour pressure of the water in the cell sap. Any such reduction in the vapour pressure of the intercellular spaces results in decreasing the steepness of the vapour pressure gradient between the intercellular spaces and the outside atmosphere, and hence decreases the rate of transpiration.

The experimental results of That indicate that the vapour pressure of the intercellular spaces is often at less than a saturation value. During the course of a day, however, the steepness of the vapour pressure gradient through the stomates may be decreased in still another and perhaps more important way. When the stomates open in the morning the vapour pressure gradient between the saturated internal leaf atmosphere and the outside atmosphere is short and hence steep, being approximately equal in length to the depth of the guard cells. For some time after the stomates open essentially such a condition is probably maintained. As the day advances the rapid loss of water-vapour molecules out of the interc-ellular spaces, perhaps coupled with a gradual reduction in the vapour pressure of the cell walls makes it less and less likely that a condition of saturation can be maintained throughout the intercellular spaces.

The zone in which a vapour pressure inequilibrium with the water in the mesophyll cell walls is maintained almostcertainly retreats more and more deeply into the intercellular spaces. Eventually it may be restricted to a thin layer just above the evapourating surfaces of the mesophyll cell walls. This gradual lengthening of the vapour pressuregradient through the stomates is probably an important factor in bringing about a reduction in the rate of transpiration during the afternoon hours.

It has frequently been observed in studies of transpiration periodicity that the transpiration rate often begins to decrease during the midday period before any appreciable change occurs in the area of the stomata1 apertures. This indicates that the initial reduction in transpiration rate is induced by some other factor than a change in the diffusive capacity of the stomates. As the preceding discussion has shown this other factor is almost certainly a reduction in the steepness of the vapour pressure gradient through the stomates. By late afternoon the air temperature and the intensity of the solar radiation begin to decrease appreciably, thus inducing a decrease in the temperature of the leaf. This lowering of the leaf temperature may further decrease the vapour pressure of the intercellular spaces, and hence further depress the steepness of the vapour pressure gradient, since temperature changes have very little influence on the vapour pressure of the outside atmosphere under standard day conditions.

Thus the effect of a reduction in the vapour pressureg radient through the stomates in decreasing the rate of transpiration, which first becomes apparent during the midday period, continues with augmented effect as the hours of darkness approach. Continued diminution in the leaf water content due to the excess of transpiration over absorption eventually results also in a diminution of the turgor of the guard cells due to a decrease in the water content of the leaf.

This results in a gradual closure of the stomates. Some of the stomates on a plant probably begin to close even before the time at which the peak of the transpiration rate is attained. In all likelihood stomates near the margin or tip of a leaf begin to close before those in the middle since the effects of deficiency of water usually appear first in these regions of a leaf. With an increasing leaf water deficit more and more of the stomates on the plant close.

This soon results in a reduction in the aggregated iffusive capacity of the stomatal population of the plant. As the afternoon advances more and more of the stomates become completely closed, resulting in a progressive diminution in diffusive capacity. This gradual closing of the stomates, resulting in an increasingly larger proportion of them being closed as the day advances, is the principal factor reducing the rate of transpiration during the latter part of the day. The effect of this factor and that of a decreased steepness of the vapour pressure gradient overlap during a large part of the afternoon.

Finally, by late afternoon, complete closure of essentially all of the stomates has occurred. Stomata1 transpiration is terminated at that time and during the hours of darkness the rate of water loss from the plant is controlled by factors influencing the rate of cuticular transpiration. Under environmental conditions deviating very greatly from those which were postulated in the preceding discussion, transpiration periodicity curves may be entirely different from those. Variations in temperature, light intensity, humidity, and soil water supply may all markedly influence both the trend of transpiration periodicity, and the magnitude of the daily water loss.

Low temperatures may result in a complete elimination of stomatal transpiration by inducing stomatal closure. Low light intensities, such as those existing on cloudy days, are unfavourable to stomata1 opening in most species. The stomates seldom open completely under such conditions and the period during which they are open is usually of shorter duration than on clear days. Furthermore, the vapour pressuregradient through the stomates is seldom as steep on such days as on clear bright days, since leaf temperatures never appreciably exceed atmospheric temperatures except when exposed to direct sunlight, and atmospheric vapour pressures are usually higher during cloudy days than on clear days at the same season of the year. The magnitude of transpiration under such conditions is usually greatly reduced, and transpiration periodicity curves plotted for plants exposed to such conditions usually present a somewhat fore shortened and greatly flattened appearance.

A deficient soil water supply is most commonly the factor which causes marked departures from the type of transpiration periodicity already considered, especially during the summer months. The effect of a gradual diminution in soil water content upon transpiration periodicity. A reduction in soil water content has two pronounced effects upon the daily march of transpiration. The total daily magnitude of water loss is decreased, and the peak of the transpiration curve often occurs somewhat earlier in the day than under conditions of abundant soil water supply. Since, even in temperate regions, periods of decreased soil water supply are of common occurrence during the summer months, and in many habitats are the rule rather than the exception for many species of plants much of the time.

Internal factors may also be responsible for transpiration periodicities of a different trend from those which have already been discussed. In some species of plants, as has already been described, the stomates remain open to a greater or lesser extent during the hours of darkness. Such plants have higher transpiration rates at night than those in which the stomates are closed. In some species of cacti there is a complete inversion of the usual transpiration periodicity curve, transpiration rates being regularly greater at night than in the daytime. This appears to be due to complete or nearly complete stomatal closure during the hours of daylight, while the stomates are, as a rule, open at night.

METHOD OF WEIGHING POTTED PLANTS

This method can be employed only with plants which are rooted in pots or other suitable containers. For laboratory experiments potted plants are often used, the pot generally being enclosed in a metal shell, and the soil surface sealed off so that evapourational water loss can occur only through the plant. For field or large-scale experiments, metal receptacles have been found convenient in which case it is only necessary to seal off or cover the soil surface in such a way as to prevent evapouration.

The method is limited in practical application to plants which can be grown in readily portable containers. The transpiration rates of plants as large as mature maize plants have been measured by this method. The loss in weight of the container and plants for a given time interval may be considered to represent transpiration as the effect of other factors on the weight of the set-up is usually negligible. If the experiment is to be continued for more than a relatively short period it is necessary to provide the container with a watering tube through which known volumes of water can be introduced into and distributed throughout the enclosed soil at appropriate intervals. In employing this method the receptacle in which the plants are growing may be weighed at selected intervals by manual manipulation or may be placed on a balance which is so arranged that each loss of a definite increment of weight (for example *i.g.*) is automatically registered on a recording device.

METHOD OF COLLECTING AND WEIGHING MTRANSPIRED WATER-VAPOUR

This method requires a rather elaborate experimental set-up, but is the only way in which the rate of transpiration can be determined quantitatively for plants growing in out-of-door habitats. The plant is enclosed within a glass chamber, the soil surface sealed off, and a stream of air taken directly from the atmosphere circulated rapidly through this container. After passing through the chamber the air is conducted through tubes or vessels containing a water absorbent such as calcium chloride. The gain in weight of these absorption tubes during the experiment represents the water-vapour transpired by the plant plus the water-vapour which was introduced into the system from the outside atmosphere.

In order to determine the proportion of the water-vapour which comes from the atmosphere it is necessary to set up a check apparatus in exactly the same manner, except that no plant is enclosed in the glass chamber, and to pass air through it at the same rate that it is circulated through the set-up containing the plant. The gain in weight of the water-absorption tubes in the second apparatus represents water-vapour from the atmosphere. Such a method has been used by Minckler for measuring transpiration from attached branches of trees.

POTOMETER METHODS

Of a limited usefulness for the measurement of transpira-tional water loss are instruments known as potometers. The severed stem of a plant is immersed in water in the reservoir of the potometer and the rate of water loss determined by the rate at which the volume of water in the apparatus shrinks. This is usually followed by noting the rate of movement of an air bubble in the water in the capillary side arm of the instrument. Some potometers are so constructed that the entire root system of plants which have been specially grown for the purpose in solution cultures can be immersed in the reservoir of the instrument.

A potometer actually measures the rate of absorption of water rather than the rate of transpiration. While under many conditions the rates of these two processes are virtually equal, this is not always true, particularly if an internal water deficit exists in the plant. The rate of transpiration as measured for a cut shoot in a potometer does not necessarily bear any relation to its rate of transpiration while it was still attached to a plant. The reasons for this will become clear in the discussion of the internal water relations of plants. The principal utility of potometers is in laboratory demonstrations of the effects of various environmental factors upon the rate of transpiration.

HYGROMETRIC PAPER METLZODS

When filter paper is impregnated with a dilute (about 3 per cent) solution of cobalt chloride and dried it becomes a bright blue in colour. If exposed to moist air its colour gradually changes to pink. The same colour change ensues when a piece of such paper is brought into contact with the transpiring surface of a leaf. If small pieces of such paper are so mounted as to be protected from the water-vapour of the atmosphere by glass, mica, or celluloid, and are brought into contact with the surface of a leaf the rate at which the paper changes in colour from blue to pinkish an indication of the rate at which water-vapour is being lost by that leaf.

A leaf which changes the colour of a piece of cobalt chloride from its full blue colour to its full pink colour in, for example, 30 seconds is losing water to the paper twice as fast as a leaf which requires 6o seconds to accomplish a similar change. This method gives no measure of absolute rates of transpiration because when a portion of a leaf is covered with a piece of cobalt chloride paper the environmental conditions influencing the leaf under the paper are very different from those which would influence it were it freely exposed to the atmosphere. The leaf under the paper is exposed to a edietd Iight Intensity and an initially lower vapour pressure than freely exposed parts of the same leaf, and furthermore, is completely isolated from any effects of wind.

Hence the rate of loss of water to the paper may be very different from the rate of water loss of the same area of leaf surface to the atmosphere. Under certain conditions this method can be used for a determination of the relative rates of transpiration of different species with a fair degree of accuracy. Even for relative determinations of transpiration rates, however, this method gives valid results only when all of the plants are growing under essentially the same atmospheric conditions. Modifications and limitations of this method are discussed by Livingston and Shreve and Meyer.The transpiration rates of plants may be expressed in terms of a unit of leaf surface (usually per square decimeter), per unit of fresh or dry weight of the leaves or tops of the plant, or per plant. Each of these modes of expression of transpiration rates is of value for certain purposes.

VARIATIONS IN TRANSPIRATION RATES

In temperate regions transpiration occurs predominantly during the warmer months of the year and especially during those periods of the warm season when the soil water supply is abundant. Among the deciduous group of woody perennials, the branches are defoliated during most of the autumn, all of the winter, and the earlier part of the spring.

Although the twigs and branches which remain exposed to the atmosphere are completely encased in corky layers of bark and the buds are enclosed within cutinized bud scales some water loss occurs from such species even during the winter months. Young twigs lose water under such conditions faster than older ones. Winter transpiration rates of deciduous woody plants are always negligible in comparison with summer rates. In evergreen species of either the needle-leaved or broad-leaved type the transpiration rates are usually not appreciably different at most times during the winter from the transpiration rate of deciduous trees at that season.

This is probably due principally to the fact that the low temperatures of the winter prevent opening of the stomates although the relatively low leaf water contents characteristic of this season may also be a factor in maintaining the stomates in a closed condition. Mild periods of any considerable duration often have a detrimental effect upon evergreen species during the winter months. The warm air temperatures induce stomatal opening, and a relatively high rate of transpiration ensues.

The soil is apt to be deficient in water at such times, or even if present, that in the surface layers may remain frozen during a period of warm air temperatures. Even if neither of the foregoing conditions exists, the soil temperatures will be low, and this greatly retards the rate of absorption of water.

The combined effect of a relatively high transpiration rate and relatively low absorption rate results in a gradual desiccation of the leaves and branches of the plant. This diminution in the water content of the aerial organs during warm periods in winter is more severe in windy weather and is more likely to occur if a sequence of mild days follows immediately after a cold spell. If severe enough this desiccation will result in the death of some of the branches, or in extreme cases of the entire tree or shrub. This is one cause of the phenomenon known as winter-killing. Winter-killing due to desiccation of the tissues should not be confused with cold injury due to low temperatures which is an entirely different phenomenon.

THE MOVEMENT OF WATER THROUGH THE PLANT

Most terrestrial plants obtain the water which is necessary for their existence from the soil. An overwhelmingly large proportion of the water which is absorbed by the roots of land plants is lost in the process of transpiration. Smaller quantities are utilized in growth and in photosynthesis and in some species limited amounts of water may be lost by guttation.

Water must therefore move through the intervening tissues and organs from the absorbing regions of the root to the tissues in which it is utilized, or from which it passes out of the plant. The process whereby water moves through the plant is often termed the conduction, transport, or translocation of water.

In herbaceous species and many shrubby plants the distance throughwhich water moves in passing from the root tips to the leaves is usually not more than a few feet. Even in such plants appearances are sometimes deceptive, as some herbaceous and shrubby species such as alfalfa may have such deep root systems that some of the absorbed water often ascends for distances as great as twenty or more feet before it reaches the level of the soil surface. It is in trees, however, in which the most striking illustrations of the upward movement of water occur. The tallest tree of which we have an authentic record is a specimen of the Coast Redwood which has attained a height of 364 feet.

Many other individuals of this species and of several others, including the Big Trees of California, the Douglas Firs (Pseudotsuga mucronata) of the Pacific Northwest, and the Blue Gums (Eucalyptus) of Australia exceed 300 feet in height. Trees ranging from 100 to 200 feet in height were of common occurrence in the virgin forests of eastern North America.

Since the root systems of trees always penetrate at least a few feet into the ground the actual vertical distance through which at least a part of the water absorbed is elevated is always more than the height of the tree. In trees, therefore, water must ascend to heights ranging up to nearly 400 feet above the level of water absorption.

The mechanism by which this feat is accomplished in tall trees has been the subject of much experimentation, and even more speculation. The ensuing discussion of the ascent of sap through plants will be presented largely in terms of its movement through trees, principally because much of the experimental work on this problem has been performed on woody species. Any explanation of this phenomenon which can be shown to be adequate for tall trees should also prove satisfactory for vascular species of lesser stature.

THE PATH OF WATER THROUGH THE PLANT

Water enters the plant through the epidermal cells and root hairs at or near the tips of the roots and crosses the cortex, endodermis, and a part of the pericycle before it finally enters the lumina of the vessels or tracheids of the root xylem. At this point its upward movement begins. The xylem tissue is continuous from just back of the root tips, through the roots, into and through the stems, the petioles of leaves, and ultimately, usually only after much branching, terminates in the mesophyll of the leaf.

The xylem tissue through which the water moves is thus a continuous unit system within the body of the plant. Along most of its course water moves en masse through the vessels or tracheids of the xylem. At the termination of the xylem ducts in the leaves the water passes into the adjacent mesophyll cells. In the mesophyll of the leaves it moves from cell to cell, eventually most of it being lost from the cells by ognized at least since the time of the girdling experiments of Malpighi in 1671. Girdling (or ringing) a stem, so that all of the tissues external tothe xylem are removed, does not prevent movement of water to organs attached to that stem above the ring. On the contrary cutting through the xylem tissue of a stem results in almost immediate wilting of leaves attached to the stem above the ring.

ANATOMY OF STEMS

Since the mechanism of the movement of water can scarcely be understood intelligently without some knowledge of the anatomy of the tissues through which it moves, a brief review of stem structure is desirable before proceeding further with a discussion of this process.

Stems vary greatly in their structure, every species possessing some anatomical features which are peculiar to itself. Nevertheless certain general patterns of tissue arrangement have been found to prevail, and the stem structure of most species will approximate one or another of these general arrangements.

The corn stem represents an herbaceous monocot type of structure while that of the tulip poplar is typical of woody dicot stems. The structure of woody stems cannot properly be appreciated merely from a consideration of one year old stems. The primary tissues of such stems develop from tissues formed at the apical growing tip during growth of the stem in length. Nearly all perennial stems also grow in diameter as a result of the development of secondary tissues from the cambium.

By successive stages of division, enlargement, and differentiation of cambium cells additional layers of xylem (secondary xylem) are laid down on the inner face of the cambium, and new layers of phloem tissue (secondary plzloem) on its outer face Hence after a few years the great bulk of any woody stem is composed of secondary tissues. Secondary xylem and secondary phloem also develop from the cambium in most herbaceous dicot stems but in such species the cambium in any one of phloem and xylem elements from the cambium.

- A cambium initial,
- Division of cambium cell,
- Outer cell resulting from division becomes a phloem element,
- Another division of the cambium cell,
- Outer cell resulting from second division becomes a xylem element.

Actually several xylem or several phloem elements are often formed successively. Upon maturation most of the xylem and phloem elements acquire sizes and shapes which are very different from those of the cambium initials from which they originate. Stem is never active for more than one growing season. The spring formed xylem tissue, as viewed in cross section is usually distinctly different in aspect from that formed later in the season.

In many angiosperms the springwood contains more and larger vessels and the cell walls are generally thinner than in the subsequently produced summer wood. In the conifers the spring formed tracheids are thinner-walled and of larger cross-sectional diameter than those formed later in the growing season. The transition from spring to summer wood is often a very gradual one. On the other hand the more open xylem tissues formed each spring abut directly upon the denser tissues which were produced during the preceding summer, thus giving rise to an abrupt line of demarcation between the zones of xylem formed in any two successive seasons.

The result of this growth behaviour is that a cross section of the trunk or a branch of any tree appears as a system of concentric layers, the so-called annual rings, each representing an annual increment of growth. In rare cases no annual rings nor more than one annual ring may be produced in a season, but usually each ring represents the xylem produced by the activity of the cambium during one season.In many woody species as the xylem tissues increase in age important changes occur in the colour, composition, and structure of the various elements, resulting in the conversion of sapwood into heartwood. As sapwood ripens into heartwood the walls of any remaining living cells of the xylem become increasingly lignified death of these cells soon following. The water content of the tissues is generally reduced and such compounds as oils, resins gums, and tannins accumulate in the cells or cell walls. The darker colouration of the heartwood of most species as compared to the sapwood is due to such accumulations. In mature trees the heartwood becomes merely a central supporting column surrounded by a cylinder of sapwood which varies in thickness from a few to many annual layers, depending upon the species and the environmental conditions under which the tree is growing.

In some species (apple, elm) the heartwood remains virtually saturated Tangential section of a small portion of the wood of whitepine (Pinus strobus). The vertically oriented elongated elements are tracheids. Bordered pits show in sectional view in the tracheid and marginal ray cells. Simple pits show in sectional view in other ray cells. Half-bordered pits show between marginal ray cells and other ray cells.

The water in the heartwood of such species as apple and elm appears to be largely static and is not directly involved in translocation. Coincident with the development of secondary xylem, secondary phloem tissues also develop from the cambium. Cork cambiums are also initiated in the bark which produce cork layers. Profound modifications therefore occur in the outer tissues as well as in the xylem of woody stems as they grow older.

A somewhat more detailed concept of the structure of the cells and elements of the stems of angiosperms through which the movement of water occurs is presented which represent longitudinal tangential and radial sections from the xylem of a tulip tree which may be taken as representative for this group of plants. The xylem tissues of the wood of angiosperms are composed of vessels tracheids, fibres of several types, wood parenchyma, and xylem ray cells. There is a tremendous variability in the proportional distribution and arrangement of these tissues according to species.

The most characteristic elements of the xylem tissue of angiosperms are the vessels. These are, in general, more or less tubular structures which may extend through many feet of the xylem. In some species cross walls, usually perforated, are of frequent occurrence in vessels, in others such cross walls are infrequent or lacking. In diameter they may range in trees from about 20 to about 400. In vines they may be as much as 700 in diameter. The vessels branch extensively in certain regions of the plant, especially at nodes within the leaf lamina, and at the points in the root systems where rool branching occurs.

The vessels of the proto xylem (the first cells of the xylem to mature during the ontogeny of a growing stem or root tip) have cellulose walls which are reinforced by distinctive lignified thickenings which appear as rings, spirals,or other characteristic patterns. The vessels that develop later in the ontogeny of any growing tip have lignified, usually pitted walls which lack any of the thickenings which distinguish the walls of the vessels of the protoxylem.

Pits normally occur in all parts of vessel walls which are contiguous with other vessels or cells. These pits consist essentially of thin areas in the walls. The walls of the vast majority of plant cells are pitted. The architecturally distinct type known as the bordered pit.

Tracizeids are found in wood of many but not all species of angiosperms. They are typically more or less spindle-shaped cells with thick walls which are almost always lignified. As viewed in cross section they are usually angular. Mature tracheids contain no protoplasm and hence are non-living like the xylem vessels. The largest tracheids are about 5 mm. in length, and about 30 in diameter, although most of them are smaller. The walls of tracheids are pitted like those of vessels. Tracheids are water-conducting cells but in most angiosperms they are relatively of much less importance as channels throughwhich water moves than the vessels.

Tracheids are developed from the cambium by a process essentially similar to that by which vessel segments are formed, except that the size and shape of the resulting cells is different, and that there is no coalescence of the individual elements such as occurs in the formation of vessels. The xylem of the angiosperms also contains fibres of various types, which in general are similar to the tracheids in structure except that usually they have thicker walls, smaller lumina, and fewer and smaller pits. Fibre cells are non-living and their walls are lignified. Because of their thick walls and small lumina the fibres do not play an important part in the movement of water through stems.

The xylem tissue of practically all angiosperms also contains wood parenchyma cells which, unlike the elements already described, remain alive for some time after their differentiation. Usually death of the wood parenchyma cells does not occur until the wood of which they are a part is converted into heartwood. Wood parenchyma cells are generally somewhat elongate, and roughly four-angled in cross section. They occur in the xylem as vertical series of cells placed end to end. The strands of wood parenchyma thus produced often extend for a long distance in a vertical direction through the wood.

The distribution of the wood parenchyma strands throughout an annual ring of xylem tissue varies with the species. In some they are more or less scattered through the xylem, in others they occur only in the last layer or two of cells which are produced in the summer woodin other words at the termination of the seasons growth and in others only in contact with the vessels, or in contact with other wood parenchyma cells which are themselves in contact with the vessels. All of the xylem elements previously described are oriented with their long axes in a vertical direction.

In addition to this vertical system there is also present in the xylem a transverse, radiating system in which the long axis of the cells is at right angles to the long axis of the stem. These transversely oriented tissue units are known as vascular rays. In the stems of most species they are continuous from the outer extremity of the phloem through the cambium into the xylem, which they penetrate to a greater or lesser distance. The portion of the vascular ray found in the xylem is termed the xylem ray; that found in the phloem the pliloem ray. Xylem rays may vary from one to many cells in thickness and likewise in height, certain types usually being characteristic of any one species. The cells of the xylem rays,like those of the wood parenchyma, usually remain alive until the woody tissue is converted into heartwood. The cells of the xylem ray are typically elongate and more or less angular in cross section.

The xylem rays probably serve as routes along which lateral movement of water occurs from the xylem to the cambium and phloem, and along which translocation of soluble foods takes place from the phloem to the living cells of the xylem. The living ray cells are in contact at various points with the strands of living wood parenchyma cells. The vertically oriented wood parenchyma strands and transversely oriented xylem ray strands thus form a unit system of living cells within the woody cylinder.

Hence there is present a continuous inter meshing network of living cells throughout the greater mass of non-living vessels, tracheids, and fibres in the younger portions of any woody angiosperm stem. There are probably few if any of the conducting elements vessels and tracheids which are not in contact at one or more points with this continuous system of living cells. The wood of the gymnosperms is simpler in its structure than that of the angiosperms, this group of woody species also showing, in general, a greater uniformity in stem structure than the latter group. The only cell types universally present in the wood of coniferous trees are the tracheids and woodray cells. Wood parenchyma cells are also present in the wood of most species of conifers, while in many species tracheid-like fibre cells are also present. The most important distinction between the wood of conifers and that of angiosperms is the total absence of vessels in the former.

The tracheids are the distinctive element in conifer wood, constituting as they do the great bulk of all the woody tissues present in such species. In conifers the tracheids form a densely packed type of woody tissue, composed of interlocking cells. Vertically contiguous tracheids always overlap along their tapering portions. Movement of water and solutes from one tracheid to another is facilitated by means of the bordered pits in the adjacent walls. Because of the numerous cross walls in a xylem tissue composed almost entirely of tracheids, water encounters a greater resistance in moving through such tissues than in traversing woody tissues which contain vessels. Nevertheless it is interesting to note that the tallest trees in the world are conifers in which all movement of water occurs through tracheids.

Theories of the Mechanism of the Movement of Water through Plants. A number of different theories of the mechanism by which the ascent of water is brought about iii plants have been suggested, and it is probable that more than one mechanism is involved in this process.

The present state of our knowledge justifies a discussion of only three possible mechanisms:

1. That ascent of sap is caused by the vital activities of living cells, principally those in the stem,
2. That this process occurs according to a mechanism described in the cohesion of water theory, and
3. That upward movement of water occurs as a result of root pressures. In passing, it should be noted that neither atmospheric pressure nor capillarity can account for the rise of sap in plants, although both of these processes are recurringly suggested as being involved in this process. Atmospheric pressure could never account fora rise of more than about 10 metres, and most mature trees attain a greater height than this. The capillary rise of water in tubes 10 in diameter is only about 3 metres, and few if any of the conducting elements have as small a diameter as this.

The mechanism of the movement of water through the plant is inextricably bound up with the mechanism of the absorption of water, so the following discussion will necessarily treat to some extent of both of these processes. Vital Theories. While the vessels and tracheids through which longitudinal transit of water occurs are non-living, they are always in more or less intimate contact with living cells. Suggestions have therefore been made from time to time that the motive power causing movement of water is furnished by the living cells of the stem. The most recent advocate of such a theory is Bose. As an example of the vital theories of the ascent of sap we may consider the proposals advanced by Godlewski in 1884. He considered that conduction of water occurred as a result of periodic changes in the osmotic pressure of the living cells present in the xylem (especially the wooJ ray cells). Upon increase in their osmotic pressure water was supposed to move into the cells from bordering xylem elements.

Upon decrease in osmotic pressure water was supposed to move back into a xylem duct. In the xylem duct it was supposed that water would move upward until it came in contact with another living ray cell because of a lower air pressure in the upper than in the lower part of the vessels. The ascent of water in stems was thus supposed to be brought about by a number of repetitions of this same process. There is little or no direct evidence in support of this or any of a number of more or less similar theories which have been suggested.

The experiments of Strasburger demonstrated quite clearly that the primary mechanism of the rise of sap in trees operates independently of the living cells of the stem. He performed many experiments upon the movement of water through woody stems in which the living cells had been killed by one method or another. In one experiment, for example, he used a 75 -year - old oak tree about 22 metres in height. This was sawed off close tothe ground and the cut end transferred to a solution of picric acid, which is toxic to living cells.

The picric acid solution slowly moved up the stem. Fuchsin, added to the liquid in which the base of the tree was immersed 3 days after the picric acid, also ascended to the top of the tree through tissues in which the living cells had been killed by the picric acid. Water also continued to ascend through similar stems after they had been completely killed by exposure to a temperature of 900 C. Similar experiments have been performed by later investigators. All of these investigators have confirmed Strasburgers results that water would continue to ascend for some time through stems, segments of which had been exposed to one treatment or another which would kill all living cells present. In all experiments of the type just described, it has been observed that the leaves at the top of a stem, part or all of which has been killed, sooner or later wilt and wither, although this effect is by no means an immediate one,often appearing only after several days. Proponents of the vital theories have accepted this as evidence that the living cells of a stem are essential for the passage of water through it. Dixon however, considers that this delayed lethal effect on the leaves is due to either or both of two causes.

Killing of the stem tissues often causes the formation of substances which plug up the vessels or tracheids, thus impeding the upward movement of water. Furthermore, death of the cells in the treated regions causes the release into the conducting channels of toxic compounds which, when transported to the eaves, cause death of the leaf cells. There is still some doubt, however,whether all of the retardation in rate of translocation as a result of such experiments can be explained in these two ways. The possibility remains open that the living cells of the xylem may in some way be necessary to the maintenance of the water-conducting capacity of the xylem, or even that they may contribute in some way to the lifting of water through stems.

In general, the evidence appears to be quite conclusive that the living cells of the stem do not furnish the principal mechanism which causes the upward transport of water in stems. On the contrary, as shown in the subsequent discussion, the living leaves or stem tips at the top of the plant appear to be absolutely necessary for the ascent of water through stems. Root Pressure. That a slow exudation of sap often occurs from the cut surface of a stem is well known. Such sap exudations are often ascribed to the development of a pressure in the dilute sap of the xylem vessels resulting from the operation of some as yet not fully understood mechanism in the root cells. Hence the name root pressure. It is generally considered that such pressures may occur in intact plants as well as in those which have been cut into. It is probable that not all of the phenomena which have been classed under the name of root pressure are due to the operation of the same physiological mechanism. Some sap exudations seem to come from the cambium or phloem rather than from the xylem. It seems quite certain that some of the sap exudations from the xylem occur from the living wood parenchyma or woodray cells rather than from the vessels, or at least that the exudation is greatly influenced by the activity of such cells.

Hence the living cells of the stemas well as those of the root are probably involved in many root pressure phenomena. James and Baker go so far as to regard the subjection of the sap in vessels to a pressure as an exceptional occurrence in plants, but this seems to be an extreme point of view, and will not be followed in this discussion. Because of the unsettled state of our knowledge regarding this phenomenon it will be necessary to use the term root pressure in a very uncritical sense. The magnitude of root pressures is commonly determined by attaching a manometer to the cut surface of the plant from which the bleeding is occurring. With only a few rare exceptions the exudation pressures which have been recorded for plants do not exceed 2 atmos., and most of them are less than this. According to White individual excised roots of tomato plants exude sap from their basal ends with a pressure of more than 6 atmos.

It has not yet been shown, however, that such pressures operate for any considerable distance through plants. In general there is no correlation between the pressure with which the sap is exuded from cut stems and the volume of sap flow. In some species relatively large volumes of sap may be exuded under a relatively low pressure, in others exactly the opposite relation may exist. Some of the woody plants in which bleeding from cut surfaces occurs,commonly ascribed in part or entirely to root pressure, are the sugar maple,box elder, dogwood, hornbeam (Ostrya), sycamore, birches, currant, and grape. Such exudation of sap under pressure usually occurs only during the seasons of the year when the plant bears no foliage, and most commonly in the early spring months.

The popular concept of a rise of sap in woody plants in the spring is largely based on the observation of such phenomena. Exudation of sap under pressure can also be demonstrated in many species of herbaceous plants especially if potted plants are used and the soil is flooded with water before the demonstration is attempted. The sap which moves through woody plants in the spring and, less abundantly, in the autumn under the influence of root pressures contains appreciable concentrations of carbohydrates as well as traces of mineral salts. In contrast with this is the condition of the xylem sap in the summer at which season it contains traces of mineral salts, but practically no soluble carbohydrates.

It has sometimes been claimed that root pressures are adequate to account for the rise of water in herbaceous plants and in low shrubs or trees,and that they may even play a very considerable part in the ascent of sap in taller trees. While it is undoubtedly true that root pressure does, in some species of plants, under certain conditions, account for the movement of some water in an upward direction through plants, this process is inadequate to account for the translocation of any considerable amount of water through plants. In the first place there are many species in which this phenomenon has never been observed.

In the second place, the magnitude of the pressure developed is seldom sufficient to force water to the top of any except relatively low-growing species of plants. Neither is the rate of flow, as it occurs inmost species, anywhere near rapid enough to compensate for the known rates of transpiration. Finally, and this is perhaps the most fundamental objection to the idea that root pressures play a very prominent part in the ascent of sap in plants; they are usually negligible, in temperate regions at least,during the summer period when transpiration is most rapid. During periods of rapid transpiration, the cut surfaces of plants not only fail to exude sap, but will usually absorb water if it is supplied at the cut surface.

The trunks of many species of deciduous trees exhibit a marked seasonal variation in water content. A minimum water content is attained during middle or late summer and the trunks gradually fill up during the fall and spring. Often there is a temporary shrinkage in their water content during the winter months but the main period of diminishing trunk water content occurs in the spring and summer. Some authorities even speak of a seasonal rise and fall in the water table of a tree trunk, and consider that intercellular spaces as well as cells become occupied with water during this process. Root pressure undoubtedly accounts at least in part for such seasonal replenishments of the store of water in tree trunks.

The living cells of the xylem probably play an active part in this process. It is possible that there may be slowly ascending streams of water in the living cells of the xylem which operate entirely independently of the water columns in the vessels or tracheids. Hence the same portions of the xylem which at times contain gases may at other seasons be occupied by water. The trunks of many conifers show no marked seasonal variations in water content; this is possibly correlated with the lack of appreciable root pressures in such species. The Cohesion of Water Theory. The leading advocate of this theory has been

Dixon who has performed much of the experimental work upon which it is based.

A number of other workers, notably Askenasy and Renner have also contributed important theoretical and experimental evidence in support of this theory. Molecules of water, although ceaselessly in motion, are also attracted to each other by strong cohesive forces. In masses of liquid water the existence of such cohesive forces is not obvious, but when water is confined in long tubes of small diameter the existence of attractive forces between water molecules can often be demonstrated. If the water at the top of such a tube be subjected to a pull the resulting stress will be transmitted all along the column of water, due to the mutual attraction between the molecules.

The water conducting system of plants constitutes just such a system, enclosing continuous thread-like columns of water which extend from the top to the bottom of the plant. A stress applied at any point to this system will be propagated to all its parts. The application of such a stress stretches the water into taut threads, throwing it into a condition of tension, which is the equivalent of negative pressure. According to this theory forces develop in the upper parts of plants, and especially in the leaves, which cause the rise of water through the plant. As evapouration proceeds from the walls of the mesophyll cells into the intercellular spaces the diffusion pressure deficit of the mesophyll cell walls increases. Water therefore moves into the walls from the adjacent protoplasm,this resulting in turn in the movement of water from the cell sap into the protoplasmic layer. The resulting increase in the diffusion pressure deficitof the cell sap is in turn propagated to the protoplasm and the cell sap in all parts of the cell. The osmotic pressures of the leaf cells are generally high enough that diffusion pressure deficits of sufficient magnitude to account for movement of water to the top of the plant can develop in them without a complete loss of turgor by the leaf cells.

If the cell is in direct contact with one of the branches of the xylem ducts which ramify throughout the lamina of the leaf, water will move from the vessel or tracheid into the cell this results in the development of tension in the water column terminating in this xylem element. The diffusion pressure deficit of the water in the conducting elements will be increased by the amount of this tension. Water under a tension (negative pressure) of 10 atmos., for example, has a diffusion pressure just 10 atmos. less than that of pure water at the same temperature which is not under tension. In other words its diffusion pressure deficit is 10 atmos. If the cell from which evapouration is occurring is not in direct contact with a xylem element a gradient of diffusion pressure deficits, gradually increasing in magnitude from cell to cell in the direction in which the water is moving, is established between the xylem element and the cell from which evapouration is occurring.

The tension developed in the conducting elements is transmitted along their entire length to their lower termination just back of the root tips and probably, very often at least, across the root tissues as well. The magnitude of the tension which develops in the xylem conduits is increased by conditions which favour a rate of water loss from the leaves considerably in excess of the rate at which water enters the roots from the soil. The movement of water across the cells of the root from the soil into the lower ends of the water-conducting elements must be considered as an integral part of the translocational process. At their lower terminations thexylem vessels or tracheids are in contact with the pericycle, or more rarely,directly with the endodermal cells of the root. Water moves from the adjacent root cells into the conducting elements because the tension (diffusion pressure deficit) developed in these elements exceeds the diffusion pressure deficit of the cells of the root.

When the tension in the water columns is relatively low a gradient of diffusion pressure deficits will be established across the root cells similar to those which are established in the leaf mesophyll cells when water is moving through them. The diffusion pressure deficits will increase from cell to cell along this gradient in the direction in which water is moving, *i.e.* from the periphery of the root towards the xylem.

Since, however, the osmotic pressures of root cells are usually lower than those of the mesophyll cells it is probable that the tension developed in the water columns often exceeds the highest osmotic pressures of any of the root cells through which water passes on its way from the soil to the xylem. Under such conditions the volume of water in the root cells may continue to diminish even after their turgor has been reduced to zero. As a result the walls of the cell are pulled inwards due to the adhesion between them and the contracting mass of water.

The counter pull exerted by the elastic walls of the cell will stretch the encompassed mass of water and throw it into a state of tension. Whenever a tension of sufficient magnitude has been generated in the conducting elements, therefore,the water in the root cells may also pass into a state of tension. Even under such conditions a gradient of diffusion pressure deficits would be established across the root, but the mechanism responsible for the development of this gradient would be more complicated than were no tension generated in the cells. When the water in a cell is under tension its diffusion pressure deficit is equal to its osmotic pressure plus the tension to which the water is subjected. Regardless of the exact mechanism involved, the essential fact is that the development of tensions in the water columns induces the establishment of a gradient of diffusion pressure deficits, increasing consistently from cell to cell across the root from its peripheral cell layer to the conducting elements and water will move along this gradient from the epidermal layer to thexylem tissue.

Water enters the peripheral cells of a root whenever the diffusion pressure deficit of water in the soil is less than that of the water in the epidermal cells and root hairs in the absorbing zone of the root. In most trees practically all upward movement of water occurs in the vessels or tracheids of a few of the outermost annual rings of the sapwood and in some species, especially those with ring-porous wood is almost entirely confined to the outermost ring. As is well known, in some species of trees (beech, sycamore, etc.) the heartwood of the trunk may disappear completely by decay. The fact that such hollow trees continue to live and thrive is conclusive evidence that the heartwood plays no essential role in the upward movement of water in such species.

The water-conductive system of plants should be regarded as a unit system; the individual threads or columns of water not operating separately but each as a part of a more or less complicated meshwork. The air-plugged tracheids or vessel sections become merely islands in the meshwork of intact water columns, and the capacity of the tissues as a whole to conduct water, although somewhat impaired by the presence of air in some of the conducting elements, is by no means destroyed.

6

Structure of Growing Plant Cell

INTRODUCTION

The electron microscope shows that young cell walls have a woven texture of dispersed cellulose microfibrils. The diameter of the microfibrils involved measures about 250 A. However, the electron microscope is said to be unsuited for research in cell physiology and growth because the specimens must be examined completely dried *in vacuo*.

Nevertheless, it is possible to take advantage of the results obtained by electron microscopy if they are properly combined with those of other methods. The natural state of the cellulosic framework of the living cell wall can be reconstructed from the electron micrograph if its chemical composition is known. In growing coleoptiles we have found 92.5 per cent water. If this water content is taken as an average for all cell constituents, the cell wall contains only 7.5 per cent dry matter. In corn coleoptiles 32 mm. in length twothirds of this dry material is composed of substances other than cellulose, such as pectins, hemicelluloses, wax. Therefore, the growing cell wall consists of only 2.5 per cent cellulose by weight.

As the density of crystallized cellulose is about 1.55 this is even less by volume. It shows how astonishingly loose is this framework in a growing cell wall, including plenty of space for living cytoplasm and all kinds of incrusting substances. Only when this highly hydrated, spacious network is freed from non-cellulosic substances and completely dried does it furnish the picture of a dense interwoven texture in the electron microscope. The impression of a rather compact framework is further enhanced by the high focal depth of the electron microscope, which yields a sharp picture of microfibrils situated in rather distant planes.

Whereas it is difficult to understand how a woven texture as shown by the electron micrograph can grow by intussusception, where new microfibrils can readily be interlaced. Another handicap to electron microscopy of cell walls has been the method of preparation, which previously consisted in disintegrating cells in a blendor whereby pieces thin enough for electron transmission were obtained.

From these specimens it was not possible to locate the pieces examined in the original cell wall, and the micrographs obtained did not show whether the whole wall of a cell or only parts of it are involved in growth. Muhlethaler showed, however, that the purified cell walls of macerated meristems are thin enough for direct examination without cutting them into smaller bits. This is possible because in these membranes he cellulose frame represents only about 2.5 per cent by weight of the living wall, which has a diameter of less than 1¼ in the swollen state.

So the dried cell wall freed of all incrusting substances would be only about 0.025¼ thick, if its mass were spread as a homogeneous film. Yet, according to the porous structure of the wall the dried framework is correspondingly thicker, say 0.075 ¼. This would correspond to a woven texture of about three microfibrils thick. As macerated collapsed cells consist of two superposed walls the thickness of the preparation may amount to 0.15¼, which is an appropriate value for the examination in the electron microscope. The possibility of investigating whole cells proved an opportunity for observing the same spot of the wall alternatively in the electron and in the ordinary microscope.

For this purpose the walls are stained with benzoazurin, and the resulting dichroism is examined in polarized light. As the direction of the strong absorption coincides with the main trend of the microfibrils, a mutual check on the observations in the electron microscope and the results of this indirect method is possible.

Thereby an identification of different cell types and the investigation of individual faces of a cell wall are feasible. Our first intention was to elucidate the tip growth in order to observe the formation of new microfibrils at the very tip of the cell where intuesusception growth occurs. *Root hairs* seemed a favourable object for this purpose, because their tip growth is very fast.

This research yielded the discovery that the tip of root hairs (*Zea mays*) is covered with a felt-like layer of cellulosic microfibrils. The felt is more than 1 ¼ thick and corresponds to the slime layer which covers the growth region of the hair.

The microfibrils are anchored in the interwoven cell wall. The slimy consistency of this layer is due to a large amount of interfibrillar substances with high swelling power, such as pectins. These can easily be demonstrated by pectin dyes (ruthenium red, methylene blue), whilst microchemical reactions, aiming at proving the presence of cellulose, fail.

This is probably due to the fact that the microfibrils found are so well crystallized that no intermicellar spaces are left for the penetration of cellulose dyes. The tip of the root hair is so woolly on account of this extracellular felt that the formation of new microfibrils cannot be observed. It is probable, however, that the cytoplasm which synthesizes the wall substances soaks this system of cellulose microfibrils, which on account of the hydrated state of the pectins must be very loose. As the process of neoformation of microfibrils is hidden in the root hairs, we looked for objects where the growth zone is not covered by slimes, such as hyphae of fungi and pollen tubes. A favourable object is the growing *sporangiophore of Phycomyces* before the tip is inflated for the formation of the sporangium. The chitinous cell walls of fungi have the same microfibrillar structure as those of higher plants; the only difference is that the microfibrils consist of chitin instead of cellulose. Curiously enough both types of microfibrils have approximately the same diameter and both are arranged in loosely woven dispersed or compact parallel textures.

At the tip of the young sporangiophore the texture of the microfibrils is very dense, and is somewhat obscured by some incrusting substance which is difficult to remove. The same difficulty is encountered with the tip of *pollen tubes*. These are very smooth and do not show any structure visible in the electron microscope.

Only after extraction with such a strong leaching agent as 10 per cent KOH can the microfibrils be made visible. Therefore, the new wall formed at the tip seems to be cutinized instantly in order to prevent desiccation of the pollen tube which is growing in air. The microfibrils of pollen tubes have a smaller diameter than those of the microfibrils found hitherto in primary cell walls. The supposition that this might be due to the haploid state of the pollen tubes turned out to be wrong. There is no relationship between the size of microfibrils and the number of genoms in the nucleus.

This was shown by looking at the haploid alga *Spirogyra,* where very coarse microfibrils occur in the cell wall. Incidentally it was found that the rounded wall of the end cell in a *Spirogyra* thread showed an unexpected tip growth. At the very tip, longitudinal microfibrils spread out from the woven texture of the wall into the open, leaving a gap between them. Along the margin of this gap cross-fibrils are woven into the pattern of the cell wall. It seems that the cytoplasm of the cell tip is so to speak naked while forming the new wall.

CELLS OF MACERATED COLEOPTILES

The extension growth has been studied in the cells of macerated coleoptiles of oat and maize. This classical subject has yielded quite unexpected results. At a very young stage (oat coleoptiles of 5 mm. length) the *parenchyma cells* show an unexpectedly early formation of local secondary bands with parallel texture laid down on the woven primary wall. These reinforcements were found to be located along the edges of the polyhedral cells.

It is difficult to understand how such a parallel-textured strip can extend when the cell elongates. These strips are also visible in the polarizing microscope when stained with benzoazurin. Between the reinforced edges there are thin pitted areas. There again an extension seems quite impossible without distortion of the pits whose elliptic orifices have always a transverse orientation. The elongation of the cells is actually not due to a growth in area of the cell faces already formed, but to a polar or even bipolar tip growth.

Since cells are rare in the macerated tissue, this tip growth must go on very rapidly. A gap bordered by longitudinal microfibrils is clearly seen, as well as a margin where transverse microfibrils are woven into this warp. In the living state the tip seems to consist of cytoplasm only, so that the wallbuilding protoplasm protrudes beyond the formed solid wall. This observation excludes any active help of the turgor pressure in cell elongation, because this would force the cytoplasm out of the cell rather than extend the woven framework of the cell wall. Cells with tip growth burst at the very tip when their turgor pressure is increased.

This impressive phenomenon of plasmoptysis at the cell tip can easily be observed when growing pollen tubes are immersed in water. *Epidermal cells* prove still more conclusively that cell extension is really tip growth. In oat coleoptiles these cells elongate from about 13 to 2000 ¼ (2 mm.), *i.e.* 150 times within 4 days.

There is a very thick exterior wall which already exists before the cell extension starts and which holds the same thickness when the extension proceeds. The thick outer wall is visible as a broad parallel-textured band at the top of the figure, whilst the rest of the cell is tubular-textured as in the parenchyma cells. At the cell tip this tubular texture grows, and the same manner of growth is used for the elongation of the thick parallel-textured exterior wall.

There a much denser tuft of microfibrils is seen which is formed by the living cytoplasm from the cell. Thus the wall lengthens in its definite thickness by bipolar addition of parallelized microfibrils. *Sieve tubes* and *tracheids* also extend by tip growth.

For young fibres and tracheids this was already known. Whilst the cell has already differentiated to a ring tracheid at the bottom of the picture, the tip consists of a veil of interwoven cellulose microfibrils ending at the top in an open tuft of the longitudinal fibrils. It is clearly seen that the cytoplasm forms the warp first, and only afterwards is the weft plaited into the woven primary wall.

Finally, the transverse microfibrils exceed those of the longitudinal warp in number. By this type of growth, parenchyma cells, epidermal cells (apart from the outer wall), sieve tubes and the primary wall of tracheids obtain their *tubular texture* with transverse pits and the highest value of anisotropic properties, such as refraction, absorption, strength, etc., in the transverse direction as had been previously established by indirect methods such as double refraction and dichroism.

STRUCTURE OF PLANT CELL

To depict the structure of a plant cell, an epidermal cell of an onion will be used as an example. The epidermis is the final tissue that covers all organs above ground. The cells of the onion epidermis are common specimens on the first day of a German basic botanical course. Since they contain no chlorophyll, they are actually no "typical" plant cells. The picture above shows an onion's epidermal cells. They are elongated and the ratio of length to width can vary strongly. Each cell is enclosed by a wall. In the region of the cell poles and where three cells adjoin, large intercellular spaces can be observed.

Elsewhere a pectin-containing middle lamina cements neighbouring cells together like bricks. The cell wall is perforated at regular intervalls, so that adjoining cells are in contact. The holes of the perforation are called simple pits and the plasma cords that run through them plasmodesmata. The surface of the epidermal cells seems to be folded, an effect that is caused by the water-repellent, waxy cuticle.

PROTOPLASM, CYTOPLASM AND CYTOSOL: THE CELL'S CONTENT

The "living" content of a cell, the protoplasm, is surrounded by a membrane called plasma membrane or plasmalemma. The protoplasm is usually next to the cell wall, so that the plasmalemma can hardly be seen. To display it, the cells are transferred into a high salt or sugar solution. As a result the protoplasm shrinks and detaches itself from the wall. The process is reversible and is called plasmolysis.

This behaviour is due to the properties of the membrane and the plasma. It is reviewed in more detail elsewhere. A substance that causes plasmolysis is called plasmolyticum and - depending on its chemical composition (potassium ions orcalcium ions, for example)—the protoplasm takes on different shapes. The plasmolyticum has accordingly an influence on the properties of the membrane. The properties of the plasma membrane differ from that of the tonoplast. The tonoplast is the membrane that surrounds the vacuole. The difference is especially striking if cells with a coloured vacuole content are used. Often the vacuole is criss-crossed by numerous plasma cords. The plasma cannot therefore not simply be viewed as a solution that is influenced by the rules of hydrodynamics alone. Rather, it contains viscous, structure-determining components, whose chemical, physicochemical and structural properties have only been recognised recently and in fragments.

Cytoplasm and Caryoplasm

The nucleus is a rather conspicuous part of nearly every living plant cell. Its structure separated from the rest of the cell by the nuclear envelope, a membrane system that consists of two discrete membranes as can be seen on electromicroscopic images. The nuclear content is called the caryoplasm while that of the rest of the cell is called cytoplasm. But these terms are only valid at certain stages of a cell's life cycle. In the course of cell-division and mitosis, the nuclear envelope disintegrates and the nucleus is replaced by the chromosomes.

It makes consequently no sense to speak of caryo- and cytoplasm during these stages. The nucleus of plant cells is usually of a round or elliptic appearance, sometimes it is also shaped like a spindle. One nucleus per cell is the rule, but cells with two or more nuclei are no rare exception. The cells of certain algae of the genus *Chladophora* have many nuclei, they are polyenergid. The nucleoli that can often be perceived after staining are substructures of the nucleus. They, too, disintegrate during cell division and mitosis and do not reshape before a new nucleus has been formed.

Plastids

Plastids are organelles that occur only in plants. Their most prominent members are the chloroplasts. Others plastids are the coloured chromoplasts and the colourless leucoplasts as well as their proplastids. Proplastidsare vestigial bodies that are generated during germ cell development due to degeneration of plastids, for example. They may differentiate into complete plastids during the development of the plant embryo. Their ripening into chloroplasts occurs usually only after light exposure.

Chloroplasts contain the green plant colour chlorophyll. They are the places where photosynthesis takes place. Chloroplasts enable the plants to convert solar energy into chemical energy. Because of this process, plants are called primary producers. The existence of consumers, like most animals, depends on them. Chloroplasts occur in most cell types, but only in organs above ground.

They can be especially well observed in tissues consisting of a single layer as in the leaflike structures of some mosses (*Funaria hygrometrica* or *Mnium hornum*) or in the water plant *Vallisneria*). Here they are rather large and of a lens-shaped appearance. During daytime, when the light is diffuse, they occur mainly at the upper and lower surface of the chloroplast. They appear to be round under top view. If exposed to strong light, they gather in parallel to the lateral sides of the cell which gives them an elleptic appearance upon top view. Chloroplasts are the site of starch production and -storage. Starch can easily be detected with the aid of potassium iodide (LUGOL's reagent). The starch-iodine complex is deeply blue-violet.

Starch production during photosynthesis can be made visible by placing a mask at a leaf that covers it partially while leaving some places exposed to sunlight. After one day of exposure, the leaf is first bleached to get rid of other pigments and afterwards treated with potassium iodide. An image is gained that is the exact replication of the mask and at the same time represents starch synthesis in the leaf.

This experiment has first been done by J. v SACHS, probably the most outstanding plant physiologist of the 20th century. He thought that starch was the primary product of photosynthesis. This assumption proved wrong. It is well-known today that the first products of photosynthesis are monomeric sugars (glucose and others) and that only part of them is used for starch production. The structures of the chloroplasts of higher cells resemble largely that of mosses. Their average diameter is 4 - 8 mm, an average cells contains 10 through 50. Their chlorophyll is unevenly distributed. At high resolutions chlorophyll-rich and chlorophyll-poor areas can be distinguished. This is due to the inner structure of the chloroplast: it is organised into grana (chlorophyll-rich) and stroma (chlorophyll-poor). Upon stimulation with short-waved light (blue or violet) the chlorophyll emits an intensive red autofluorescence that looks especially impressive in a fluorescence microscope. The differences between grana and stroma become very obvious. The uniformity of the chloroplasts of all higher plants points out that the optimal form has been found rather early in evolution and has not been changed since. This is different with algae. The chloroplasts of green algae (Chlorophyceens) are very varied in shape. Many species have just one chloroplast that covers nearly the whole space of the cell's interior. It is screw-like in *Spirogyra*-species, star-shaped in *Zygnema* and *Zygnemopsis* and netlike in *Oedogonium*.

The disc-shaped chloroplast of *Mougeotia* can be viewed either from above or in profile depending on the amount of light used. Its rotation is a well-analysed example of an induced chloroplast movement. The chloroplasts of many species of algae contain often well-visible pyrenoids, structures, that produce and structure starch. Chromoplasts are red, orange or yellow plastids. The colour is usually the result of yellow xantophyll and yellow to red carotinoids. Both compounds do also exist in chloroplasts, but are concealed by chlorophyll. Chlorophyll is broken down much faster than carotinoids as can be observed in the coloured leaves in autumn. Fluid transitions between chromo- and chloroplasts exist, just as between chromo- and leucoplasts. Typical chromoplasts cause the orange colour of the carrot, the red colour of the ripe pimento and tomato as well as the colour of numerous flowers. Carotinoids are not very water-soluble and do therefore often crystallise within the chromoplasts. Their crystals can be disc-shaped, needle-like, jagged or sickle-like. In many cases, flower and leaf colours are caused by the coloured content of the vacuole. The colour of the vacuole and that of the plastids may lead to a mixed colour. The leaves of the copper beech, where the vacuole's content is red and that of the chloroplasts is green are a typical example. The plastids of the red and brown algae are traditionally counted among the chromoplasts although they contain chlorophyll. The green colour is concealed by the red phycoerythrin (Rhodophyceae) or the brown fucoxanthin (Phaeophyta).

Leucoplasts are common, colourless plastids. They develop from proplastids, but form no homogeneous group of their own. A part of them can differentiate into chloroplasts or chromoplasts at light exposure, while this is not due for others. The guard cells, for example, contain leucoplasts, that are permanently exposed to light without developing into chloroplasts. Leucoplasts do also occur within colourless leaves (variegated leaves) or plant parts. There exists a number of examples which show that they developed from chloroplasts that lost their ability to produce chlorophyll. There are even species, like *Neottia*, an orchid that cannot produce chlorophyll at all and are thus dependent on a parasitic or saprophytic lifestyle (saprophy is the feeding from dead organic material). A second class of leucoplasts occurs within the non-green tissues of otherwise green plants. It is especially common within roots. Though these leucoplasts are capable to become green, they do usually not since they are not exposed to light.

The leucoplasts of the calyptra (a calyptra is any hood or cap of cells protecting a plant part) contain starch and are therefore counted among the amyloplasts (starch-containing leucoplasts). They have, as is explained later, the function of statolithes, that have an important part in the perception of gravity (geotropism).

THE CELL WALL

Except for very few examples, plant cells are surrounded by a cellulose containing cell wall. It is flexible and distortable during growth, but loses its ability for distortion after growth has stopped, while a limited flexibility remains. Because of these changes, it is distinguished between primary and secondary cell walls. As we will see when talking about electron microscopic pictures of the cell wall, both forms differ mainly in the arrangement of their cellulose microfibrils. While they are unorganised within the amorphous matrix of the primary cell wall, they are organised into several ordered layers that are arranged one on top of the other at right angles in the secondary cell wall. The secondary cell walls of many cells, especially those of vascular tissues, are incrusted with strengthening material.

Two important ones are:

1. Lignin, the ground substance of wood and
2. Suberin, the ground substance of cork

CONCENTRATION OF PLANT CELLS

The hydration and viscosity of the protoplasm, the permeability of the cytoplasmic membranes, the activity of enzymes, the chemical activity of various ions in the cell, and various other physiological processes and conditions are all influenced more or less by the hydrogen ion concentration of the protoplasm and the cell sap. As soon as its significance was appreciated attempts were made to determine the hydrogen ion concentration of plant cells. Most of the earlier determinations were made on the juice pressed from plant tissues. Such a crude method provides only the roughest sort of an indication of the hydrogen ion concentrations in individual plant cells. The death of the cells and the mixing of the cell contents during the extraction process undoubtedly result in marked changes in hydrogen ion concentration. Such determinations are probably more nearly a measure of the pH of the cell sap than of the protoplasm. The values for expressed plant saps mostly fall within a range of pH 3.Q to pH 7.0 Direct measurements of the hydrogen ion concentration of the protoplasm and cell sap have been made by introducing indicator dyes directly into cells.By careful manipulation of a micropipette the dyes can be injected into the cytoplasm without penetrating into the vacuole or injected into the vacuole without penetrating into the cytoplasm.

Only non-toxic dyes should be injected into living cytoplasm for pH determinations, else the results may be invalidated by injury or death of the cytoplasm. Dyes are considered to be nontoxic or essentially so if protoplasmic streaming continues in the same way as before the injection. In this manner it is possible to determine the reaction of the cell sap and that of the cytoplasm independently. Results of the micro injection method when applied to the root hairs of the water plant Limnobiumspongia indicated a pH value for the cytoplasm of 6.9 1 0.2. The cell sap of the same cells was found to be more acid having a pH of 5.2 1 0.2.

As indicated by the results cited above the pH of the cytoplasm and the cell sap of a plant cell may be very different. The pH of the cytoplasm of plant cells appears to be fairly constant, usually falling between 6.8 and 7.0. The pH of the cell sap is usually lower than that of the cytoplasm, values between pH 5.2 and 6.2 seeming typical for most plant cells.The cell sap of some cells, however, shows a considerably more acid reaction than this, values as low as p11 0.9 being reported for species of Begonia. On the other hand alkaline values have also been found in the vacuoles of some species. The cells appears to vary more widely in hydrogen ion concentration than the cytoplasm.

CELL-WALL GROWTH IN DIVIDING CELLS

How does the cell wall behave when the cell divides? In very young coleoptiles there is still some mitotic activity. We could therefore photograph a certain number of dividing cells. The structure of the expanding cell plate is difficult to observe, because it is inside the enveloping membrane of the mother cell. But the constriction of this wall in the middle of the cell is clearly visible. Its microfibrils are partly bent inwards, where they meet the cell plate.

Thereupon they are interwoven with the microfibrils of the corresponding lamella of the double-laminated cell plate. A certain number of the microfibrils of the original longitudinal wall are left over and decorate the margin of the two daughter cells with their submicroscopic fringes. These microfibrils are gradually dissolved and disappear. This proves that existing cellulose fibrils can be hydrolysed near by the place where other microfibrils are synthesized. Whether these two antagonistic processes are catalysed by the same enzymatic system is open to discussion.

The effort of our laboratory to elucidate the mechanism of extension growth has shown that, in the object investigated, there is no difference between tip growth and 'cell extension'. This had already been postulated previously. But it was thought then that the primary wall could expand by loosening its existing framework and subsequently inserting new microfibrils.

In order to obtain a local loosening, certain microfibrils had to be dissolved by hydrolysis.

That a local 'melting down' of microfibrils is possible has been proved in this study, when we observed how the ring of microfibrils left over as fringes along the border of the new wall in divided cells gradually disappears. Thus, plasticizing of an existing cellulose frame by dissolution of microfibrils is possible. But in the observed cases of cell extension it does obviously not occur. Growth in area does not proceed by intussusception but by a terminal increment, just as the area of a cloth is enlarged by weaving its margin. In so doing the cytoplasm must somehow protrude beyond its solid case.

As the coleoptiles investigated represent the classical subject for the study of extension growth, there is some danger of generalizing these findings to all cases of growth in area of cell walls and of banishing the concept of intussusception.

Although most elongating cells probably grow in the way here described, there are still some cases which must be thoroughly investigated before a general theory of cell-wall growth can be formulated. It will be very interesting to find out how tip growth proceeds in the very fast extension growth of grass filaments.

Further, there is an enormous growth in area of the wall when the descendant of a narrow cambial cell differentiates to form a large-bored vessel member. Here no tip growth is possible, but the cell surface must increase along some longitudinal line or lines, where plasticizing of the existing wall is necessary. The same is true when a cell produces some local outgrowth (*e.g.* a root hair). At the place where this reactivated tip growth starts, the existing framework must have been loosened somehow beforehand.

Since the texture of primary walls is so spacious, there are many possibilities of rendering the wall softer by increasing the interfibrillar ('incrusting') substances capable of swelling, or by intercalating new microfibrils. This would be possible where living cytoplasm is located within the growing cell wall. Our study seems to indicate, however, that the cell wall is only soaked with living matter locally. Thus intussusception is restricted to small areas, and the cytoplasm retires very soon from the woven framework. All these considerations show that the growth of the plant cell wall is an autonomic process. Exterior forces such as turgor pressure cannot play an essential role; they are only involved in so far as they influence the activity of the cytoplasm.

The extension growth of the cells in coleoptiles is a *bipolar tip growth*. The growing cell wall is living: it must be soaked with cytoplasm which synthesizes microfibrils *in situ*. In dividing cells it bends them to meet the cell plate on which they are fastened. Therefore the growth of the cell wall is an autonomic process which is governed by the morphogenetic faculties of the living protoplast.

DIVERSE FUNCTIONS OF CELL WALLS IN PLANTS

Plants do not have specialized skeletal systems of the type found in animals; instead, the considerable strength that enables large plants such as the giant *Sequoia* redwood tree to withstand the forces of gravity is provided by the cumulative strength of the walls that surround individual cells. But the load-bearing walls also have to be sufficiently flexible to withstand the large tensile and compressive forces that occur as, for example, the plant sways in the breeze. The wall is subjected to further forces from within, where osmotic pressure represents another threat to cell and plant integrity. In young cells, the walls must be porous, to allow water, low–molecular weight nutrients and hormones to diffuse freely between cells and tissues, whereas in specialized vascular tissues required for long-distance movement of water and nutrients through the plant, the walls must be impermeable.

In most tissues, cell walls are also important for intercellular adhesion. These various functional requirements are accommodated through the formation of reinforced gel structures consisting of cellulosic microfibrils, which have a high tensile strength, embedded in a gel-like matrix phase that consists predominantly of non-cellulosic polysaccharides.

This matrix provides flexibility and some mechanical support for the cell, but its inherent porosity also allows water and other small molecules to diffuse through the wall.

As cells of higher plants stop growing and increasingly assume a structural role, their walls thicken, cellulose microfibrils become laminated to increase wall strength and the walls often become encrusted with lignin, which is a large polyphenolic molecule that renders the wall resistant to compressive forces and restricts the passage of small molecules. At the same time, the wall will 'lock in' the final shape of the cell, and many different cell shapes can be detected in a single section of a plant organ. Smaller amounts of structural and enzymatic proteins and various proteoglycans and glycoproteins are also found in the wall, but these will not be discussed in detail here.

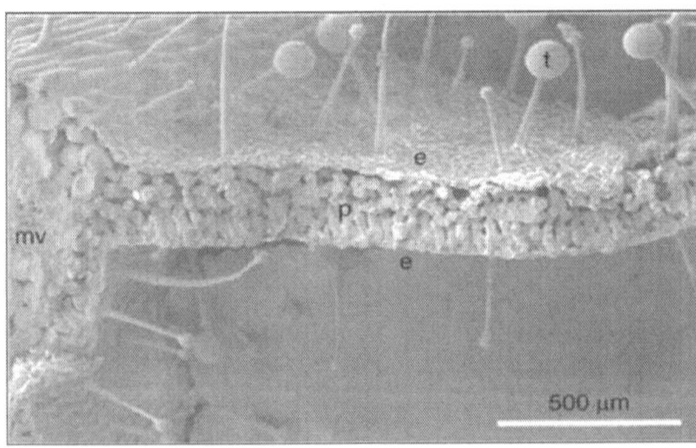

Fig. Heterogeneity in Cell Shape in a Leaf from *Nicotiana Benthamiana*.

In this mature leaf, which has stopped expanding, there are parenchyma cells (p) that have thin walls, fibre cells that have thickened walls, xylem cells that are no longer living and need a wall impervious to the water they transport, and trichome cells (t) that are greatly elongated and require a stiff wall to support them above the surface of the leaf. The cell walls surrounding these cell types play a major role in 'locking in' the final shape of the cell. The midvein (mv) and epidermal layers (e) are also indicated.

In addition to their crucial structural role, cell walls are likely to represent a significant source of metabolizable energy for some plants and may be subject to continual turnover in certain tissues as a result of this function. For example, the walls of the storage cells of seeds are not load bearing but in some cases are extraordinarily thick. Wall polysaccharides in these seeds act as an alternative to starch as a storage macromolecule. Examples include the galactomannans of seeds like fenugreek, carob, ivory nut and guar, as well as pectins in lupin seeds and xyloglucans in nasturtium and tamarind seeds. Grains of many grasses, particularly the economically important cereals such as wheat, maize, rice and barley, contain an endosperm that is rich in starch but also comprises a significant amount of cell wall material that can be used as a source of energy by the germinated grain. Indeed, it is estimated that up to 18 per cent of metabolizable glucose in the germinated barley grain is released from the (1,3;1,4)-α-D-glucans of the starchy endosperm cell wall. A similar energy reserve function has been proposed for cell wall (1,3;1,4)-α-D-glucans in vegetative tissues of the grasses.

Finally, fragments of non-cellulosic polysaccharides of cell walls have been implicated in growth regulation and in signaling pathways associated with a loss of cell wall integrity. Various oligogalacturonan fragments from pectic polysaccharides are bound by receptor-like kinases and can initiate a range of cellular changes, including defence responses that are executed following pathogen attack. Indeed, one might reasonably ask whether the evolution of the extreme complexity in pectic polysaccharides might have been driven by the potential of these polysaccharides to release an array of different signaling oligosaccharides. If this were true, an obvious next question would be whether other wall polysaccharides, such as the heteroxylans, undertake a similar function, especially in the *Poaceae,* where levels of pectic polysaccharides in the wall are very low.

Alternative Chemistries for Non-cellulosic Polysaccharides

Two major chemical strategies have been used by plants to match non-cellulosic polysaccharide structures to their functional requirements as components of the gel-like matrix phase of walls. Structurally regular polysaccharides with extended conformations of the type commonly found in cellulose and related wall polysaccharides will tend to align and aggregate in aqueous media; the aggregates will be stabilized through the formation of intermolecular hydrogen bonding and hydrophobic interactions between sugar rings. This will reduce their solubility and enhance their ability to form gels.

So, how do plants introduce limited and possibly controllable irregularity into the non-cellulosic wall polysaccharides to limit this tendency to aggregate? In most instances the structurally regular and conformationally extended backbone or main chain that would normally align and aggregate into fibrillar or insoluble complexes is prevented from doing so through the addition of single monosaccharide substituents or short oligosaccharides along the otherwise regular backbone chain. During gel formation in the wall, the presence of some structurally regular 'junction zones' along the polysaccharide chain facilitates localized intermolecular interactions, which, in combination with 'solubilizing' non-interacting regions of the chain, allows the polysaccharide to form a gel.

An example of this chemical strategy is embodied in the xyloglucans, which consist of a (1,4)-β-glucan main chain. These 'cellulosic' backbone chains do not aggregate because they are substituted with short oligosaccharide chains at the C(O)6 of the main chain (1,4)-β-glucosyl residues. Similarly, the arabinoxylans of the *Poaceae* consist of a (1,4)-β-xylan main chain that would normally be insoluble and fibrillar in nature but is instead substituted with α-arabinosyl residues. There are also regions of the (1,4)-β-xylan backbone that carry no substituents; these will therefore be capable of intermolecular interactions that will result in limited junction zone formation and will allow gel formation in the wall. The chemical structures of galactomannans and pectic polysaccharides reflect the same strategy.

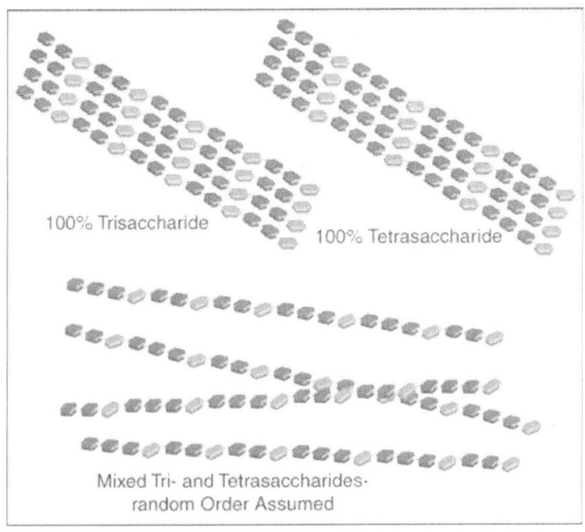

Fig. The Effects of (1,3;1,4)-β-glucan fine Structure on Aggregation and Solubility.

The second, quite different strategy to achieve wall polysaccharides with extended conformations but with limited tendencies to align and aggregate is exemplified by the relatively recently evolved (1,3;1,4)-β-glucans of the *Poaceae*. Like the xyloglucans, their structure is based upon a conformationally regular 'cellulosic' (1,4)-β-glucan backbone. However, in this alternative strategy, intermolecular aggregation is limited through the introduction of irregularly spaced (1,3)-β-linkages, which result in molecular 'kinks' in the otherwise 'cellulosic' backbone. Provided the (1,3)-β-linkages are spaced at irregular intervals, the asymmetric polysaccharide chains will not aggregate over extended regions.

The (1,3;1,4)-β-glucans of the *Poaceae* have structures based upon a conformationally regular 'cellulosic' (1,4)-β-glucan backbone. Intermolecular aggregation is limited through the introduction of irregularly spaced (1,3)-β-linkages, which result in molecular 'kinks' in the (1,4)-β-glucan backbone. The (1,3)-linked β-glucosyl residues are shown in red, whereas the (1,4)-β-glucosyl residues are shown in blue. Relatively high or relatively low ratios of cellotriosyl/cellotetraosyl units indicate a predominance of cellotriosyl or cellotetraosyl residues, respectively.

If one or another of the oligosaccharide units predominates, the overall polysaccharide chain becomes more regular in shape and hence less soluble because of chain alignment. Random arrangement of a mixture of the trisaccharide and the tetrasaccharide units means that the chains will not aggregate over extended regions and can remain in aqueous solution, as depicted in the lower diagram. Such differences in physicochemical properties could be used to the plant's advantage in meeting different functional requirements in different cells or at different stages of development.

Molecular Mechanisms for Wall Polysaccharide Biosynthesis

The generally accepted mechanism of wall polysaccharide synthesis involves enzymes that add glycosyl residues from activated sugar nucleotide donors to the elongating polymeric chain. Two major classes of these glycosyltransferases have been recognized. One group, which is often referred to as the polysaccharide synthase group, is believed to catalyze the iterative addition of glycosyl residues to the end of the elongating polysaccharide chain, without releasing the chain between successive glycosylation events. It has been generally assumed that the incoming glycosyl residue is transferred from the sugar nucleotide donor substrate to the non-reducing end of the polysaccharide, although it has also been suggested that plant cell wall polysaccharide biosynthesis might involve addition of glycosyl residues to the reducing end of the elongating chain.

The iterative synthase enzymes are assumed to catalyze the polymerization not only of cellulose but also of the main chain of non-cellulosic wall polysaccharides; these enzymes include the (1,4)-βmannan synthase that polymerizes the (1,4)-β-mannan backbone of mannans, galactomannans and galactoglucomannans and the (1,4)-β-glucan synthase that synthesizes the backbone of xyloglucans. These enzymes are encoded by members of the cellulose synthase (*CesA*) and cellulose synthase-like (*Csl*) gene superfamily and are members of the GT2 group of glycosyltransferases. The GAUT enzymes are also believed to be iterative glycosyltransferases that synthesize the (1,4)-β-galacturonan backbone of pectic polysaccharides, but the GAUT galacturonyl transferases belong to the GT8 group of enzymes, and their structures are likely to be quite distinct from the GT2 group involved in the synthesis of β-linked polysaccharide chains.

The second group of glycosyltransferases is believed to add single glycosyl substituents to the main chain residues or to add short oligosaccharide chains that are appended to the main chain through single catalytic encounters with the nascent polysaccharide chain. Examples of this group of glycosyltransferases include the arabinosyl transferases that add á-arabinofuranosyl or β-glucuronyl residues to the (1,4)-β-xylan backbone of arabinoxylans, the xylosyl transferases that add (1,6)-β-xylosyl residues to the (1,4)-β-glucan backbone of nascent xyloglucan molecules and the galactosyl transferases that add (1,6)-β-galactopyranosyl residues to the backbone of (1,4)-β-mannans or (1,4)-β-glucomannans.

Despite the best efforts of biochemists over many decades, there are still a number of fundamental questions about the biosynthetic process that we have not answered. The mechanism of chain initiation is not known, and potential primer molecules for wall polysaccharide biosynthesis have not been convincingly identified. Similarly, the mechanisms controlling chain termination and thus the overall length of the polysaccharide chain have not been elucidated.

Do polysaccharide chains simply 'fall off' the synthase enzymes when they get to a certain length or when the enzyme itself begins to be turned over? Do the hydrolytic enzymes that have been increasingly recognized in tissues in which net synthesis of wall polysaccharides is occurring hydrolyze the elongating chain and release the nascent polysaccharide from the synthase enzyme? Might decreasing concentrations of glycosyl donor substrates lead to termination of the chain?

It was recently suggested that a terminal oligosaccharide detected at the reducing end of certain heteroxylans might be involved in chain termination. Whichever of these or other possibilities might apply, plant cell wall polysaccharide synthesis does not appear to be guided by a template, and it is difficult to envision how chain termination could be anything but an imprecise process. It follows that the resultant polysaccharide chain length will be heterogeneous, as is usually observed. This is not to say, however, that plant cells do not have some control over wall polysaccharide chain length, because clear differences are apparent, and in some cases chain lengths fall within reasonably narrow limits.

A second potential level of imprecision is afforded by the glycosyltransferases that add the various glycosyl units to the backbone of the wall polysaccharide. We do not know whether the addition of the substituent is a precise process or not, insofar as there are examples of relatively regular spacing of substituents along the chain and other examples in which substituents are, in statistical terms, asymmetrically distributed. In the former case, one might expect that the backbone synthase and the substituting glycosyltransferase would have to operate as a highly coordinated enzyme complex.

On the other hand, if the main chain synthase and glycosyltransferase were operating independently, it is again difficult to see how main chain substitution could be a precise process with respect to the extent and spacing of added substituents. Equally plausible is the possibility that an enzyme complex could mediate the generation of a polysaccharide in which every main chain residue was substituted or in which every second or third main chain residue was substituted.

This would result in a relatively homogeneous polysaccharide, but postsynthetic removal of substituents at regular intervals, as has been described for arabinoxylans, or asymmetrical postsynthetic modifications such as methyl esterification, as has been described for some pectic polysaccharides, could introduce heterogeneity into a polysaccharide that was initially homogeneous with respect to its fine structure.

Against this background of heterogeneity that will arise generally as a result of biosynthetic mechanisms, the specific types of heterogeneity actually observed in the major groups of non-cellulosic wall polysaccharides are summarized in the sections below.

Heterogeneity in Pectic Polysaccharides

Pectins are a major component of the primary cell wall, particularly of dicotyledonous plants. Pectins contribute to the mechanical strength, porosity, adhesion and stiffness of the cell wall and have been implicated in many processes, including wall slippage, extension and intercellular signaling. In addition, pectic polysaccharides are a source of signaling molecules. Pectins are extremely complex molecules that include a number of structural classes, such as homogalacturonan, xylogalacturonan, apiogalacturonan and rhamnogalacturonan. The galacturonan chains can chelate calcium ions, apiosyl residues can coordinate borate ions, and there is good evidence that covalent linkages form between different structural classes of the pectic polysaccharide group. Methylated and acetylated residues are often found on pectin and indeed on several other polysaccharides and can profoundly influence their structures and physicochemical properties.

Considering this complexity, it is not unexpected that the level of heterogeneity in the pectic polysaccharides is high. Here we will explain just one example of how structural heterogeneity is generated and how this alters the physicochemical and functional properties of the cell wall matrix. A common mechanism facilitating local changes in the properties of pectic polysaccharides is the degree to which their polygalacturonate residues are methyl esterified. This involves heterogeneity both within a chain and between chains and can now be characterized analytically. Methyl esterification will neutralize the charge on galacturonyl residues and thereby abolish the ability of the negatively charged galacturonyl residues to interact with calcium ions. Specific monoclonal antibodies, which can differentially detect pectins with particular methylation profiles, show spatial heterogeneity around a single higher-plant cell with respect to the pattern of pectin esterification, which may be linked to cell adhesion.

There are a number of pectin methyl esterases that selectively remove methyl groups from homogalacturonan in either a blockwise or a non-blockwise fashion. Where methyl groups are removed in a blockwise manner, an exposed region of the polysaccharide that is able to form calcium-mediated cross-links is created; these cross-links contribute to the 'stickiness' required for cell adhesion. Where the removal of ester groups is non-blockwise, the homogalacturonan appears to be less 'sticky', and such junction zones cannot form so readily. As one might expect, this particular form of homogalacturonan is found in regions where cells separate.

Heterogeneity in Heteroxylans

Heteroxylans have a (1,4)-β-xylan backbone that can be substituted with single α-arabinofuranosyl, α-glucuronosyl or 4-methyl-α-glucuronosyl residues. These residues can be linked to the $C(O)_3$ or $C(O)_2$ atoms of the backbone xylopyranosyl residues, or in some cases the xylopyranosyl residues can be substituted at both $C(O)_2$ and $C(O)_3$. Short oligosaccharide substituents are also found on some heteroxylans and, in the grasses in particular, á-arabinofuranosyl substituents can be esterified with hydroxycinnamic acids such as ferulic acid and *p*-coumaric acid.

Two carbohydrate-binding modules and two monoclonal antibodies have been used to locate specific types of xylans both within intact cell walls and in solubilized wall material. Both classes of probe revealed distinct types of heteroxylans, which differ in both structure and charge and are present in both primary and secondary walls. Similarly, in the endosperm of the grasses, wall arabinoxylans digested with (1,4)-β-xylan endohydrolase release a population of oligomeric arabinoxylosides that provide an enzymatic fingerprint of the polysaccharide's chemical structure.

Examination of the arabinoxylan fingerprint from wheat endosperm walls showed variation in the structure and amount of this polysaccharide both spatially and temporally through grain development.

The results were confirmed through Fourier transform infrared microspectroscopy (FT-IR), which showed that arabinoxylans in developing grain contained more backbone xylosyl residues disubstituted with arabinosyl residues than did arabinoxylans in mature grain and that the structure of the arabinoxylan was dependent upon its position in either prismatic or central cells.

Heterogeneity in Xyloglucans

Xyloglucans are thought to be particularly important in the primary walls of many plants because of their ability to non-covalently cross-link cellulose microfibrils. Their structure is based on a cellulosic (1,4)-β-glucan backbone substituted at $C(O)_6$ with α-linked xylosyl residues, typically at about 70 per cent of the backbone glucosyl residues. However, further substitution of xylosyl residues is required to prevent xyloglucan self-association, which interferes with binding to cellulose. A typical basic substructure involves xylosyl substituents on three consecutive glucosyl backbone residues, followed by an unsubstituted glucosyl residue. The availability of endoglucanase enzymes that depolymerize the xyloglucan chain at unsubstituted glucosyl residues allows the isolation of oligosaccharides for detailed analysis.

A wide range of carbohydrate and selective acetyl substituents on the basic xyloglucan structure have been characterized in this way. Most studies of partial glucanase digests are consistent with an essentially random arrangement of oligosaccharide substructures, but this may not always be the case. One promising new approach to characterizing the heterogeneity of xyloglucans is the use of mass spectrometry to define glucanase-derived oligosaccharides on very small sample sizes. The selectivity of specific endoglucanase digestion, coupled with the sensitivity of mass spectrometric detection, allows not only the profiling of small amounts of isolated xyloglucan polymers but also the profiling of xyloglucans present within specific cell types or in isolated organelle fractions. Furthermore, *in situ* analysis of xyloglucan oligosaccharide substructures in walls can be carried out at high resolution by aiming the ionization laser of a MALDI-TOF mass spectrometer at desired locations within enzyme-treated tissues. Although mass spectral methods depend to some extent on prior knowledge of xyloglucan structure and cannot be easily applied to polysaccharides such as arabinoxylans, which consist only of pentose sugars, the method has potential to open up the dissection of location-specific heterogeneity in xyloglucan structures, especially if the method can be complemented with antibodies directed against an array of xyloglucan substructures.

Heterogeneity in Mannans

The cell wall polymer family for which fine structure has been most precisely defined and reconciled with biosynthesis processes is the galactomannans, members of which have a (1,4)-β-mannan backbone along which some but not all 6-C(OH) positions are substituted with α-galactose. Mannose/galactose ratios vary from about 1:1 to as high as 20:1. On the basis of enzyme digestion and quantitative analysis of digestion-resistant oligosaccharides, it has been shown that second-order Markov chain statistics are sufficient to describe the fine structure of a range of galactomannans. For a processive biosynthesis mechanism in which galactosyl transferase activity is coupled with backbone mannan synthase, a second-order Markov chain is consistent with the galactomannan subsite recognition sequence of the galactosyltransferase encompassing at least three backbone mannosyl residues—that is, the site of substitution and the two preceding residues towards the reducing end of the growing galactomannan chain. Datasets from three different plant species generated three distinctly different sets of probabilities and thus different galactosyl-substitution patterns. Subsequently, heterologously expressed galactosyl transferases in transgenic tobacco resulted in the production of galactomannan with Markov probability values characteristic of the original source of the galactosyl transferase. The Markov statistics describing polymer structure also allow a precise prediction of the probability of unsubstituted mannan segments of defined length, and these have been shown to be of consequence in the formation of interpolymeric junction zones that can form the basis for three-dimensional gel-like matrices of the type thought to be present within plant cell walls.

ORIGIN AND DEVELOPMENT OF PLANT

As soon as it became clear that all plant and animal tissues were composed of cells the question of the origin of cells naturally arose. This proved to be a difficult problem for the pioneer investigators and for many years there was much disagreement over the question. The now universally accepted principle that cells can arise only by the division of pre-existing cells was first demonstrated beyond any reasonable doubt by Niigeli just before the middle of the Nineteenth Century. A number of methods are now known by which this division is accomplished but a detailed discussion of them is beyond the scope of this book. In higher plants cell division occurs chiefly in certain restricted regions called meristems. The production of new cells involves not only the division of pre-existing cells, but the subsequent enlargement and maturation of their cell progeny.

FORMS AND SIZES

All newly formed cells do not differentiate morpholo-gically in the same way. Some elongate parallel to the axis of growth more than in other directions and thus produce the longer fibre cells that make up much of the xylem and phloem tissues. These cells commonly develop walls that are greatly thickened. Some cells enlarge about equally in all directions forming isodiarnetric cells, the walls of which are never greatly thickened. Cells of this type are present in the pith, leaf mesophyll and in other parenchymatous tissues.

According to Lewis both plant and animal cells are basically tetrakaidecahedrons; *i.e.,* 14 sided although many other geometrical shapes are found, especially in specialized types of cells. An almost endless variation in cell shapes and sizes may be seen in the tissues of any vascular plant, all of which may develop from similarly shaped meristematic cells. In size cells show an equally great range. Most plant cells have diameters that fall somewhere between 10 and 100. A single cubic centimeter oftissue may, therefore, contain millions of cells. Some cells, however, are much larger. Such cells have lengths that are many thousand times their diameters.

MITOCHONDRIA: POWER PLANT OF CELL

For thousands of years, men puzzled over the question of where body heat comes from. This heat was recognized as nearly synonymous with life itself–but what produced it? The source of the energy of man or beast was equally mysterious. In recent decades it has become clear that both heat and energy can be traced to some sausage-shaped organelles in our cells, the mitochondria. Of course the ultimate source of energy for all plants and animals is sunlight. But the sun's energy can be harnessed by plants, through photosynthesis and stored in molecules of carbohydrates.

When animals eat these carbohydrates and break them down to carbon dioxide and water, with the help of oxygen and an arsenal of enzymes, large amounts of energy become available. Animals immediately convert this energy into molecules of high-energy ATP (adenosine triphosphate) -- the universal currency of energy in living things. Excluding only the very first stages in carbohydrate breakdown, which are called glycolysis, the entire, complicated process of energy transfer to ATP takes place within the mitochondria.

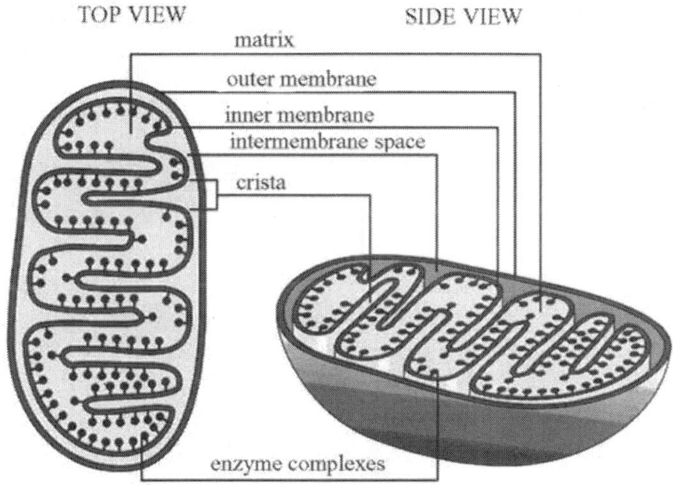

Fig. Mitochondrion.

Mitochondria are the largest organelles in the cell, after the nucleus, yet some cells have more than a thousand of them. Thin and long, they vary in diameter from 0.5 to 1 micrometer and in length up to 7 micrometers. Their thread-like outlines can be seen with a good light microscope.

Although they were first observed in 1850, it took a century to visualize their internal structure and understand their function. Actually the story goes back even further, to the 18th century, when Lavoisier showed that animals require oxygen for respiration. This led chemists to study cellular respiration, the process by which living cells use oxygen to release the chemical energy stored in foodstuffs.

Early in the 20th century biochemists discovered that these reactions fall into two major groups:

1 The carbon pathway, where a series of chemical reactions, each requiring a specific enzyme, breaks down the carbohydrates into carbon dioxide and hydrogen;

2 The hydrogen pathway, which transfers the hydrogen to oxyen in stages, forming water and releasing energy.

The whole process is organized much like an assembly line. In the hydrogen pathway, the hydrogen's electrons pass through an "electron transport chain" made up of enzymes that act as carriers and as the electrons move from enzyme to enzyme, they give up part of their energy, which is stored in molecules of ATP.

The ironcontaining enzymes, cytochrome a, b and c, carry out the final stages of the process. All in all, three molecules of ATP are formed for every atom of oxygen that is used up in respiration. For a long time, biochemists studied these reactions in cell extracts without worrying about what parts of the cells might be involved.

One scientist, B.F. Kingsbury of Cornell University, did suggest that mitochondria might be the site of cellular respiration back in 1912, but this was ignored. In the 1940's, using an ultracentrifuge, Albert Claude and his associates isolated various cell particles including mitochondria. Shortly afterwards, in 1948, Albert Lehninger of Johns Hopkins and Eugene Kennedy of Harvard showed that the reactions leading to the synthesis of ATP occurred in the mitochondria. In the early 1950's, Palade and the Swedish scientist Fritiof Sjostrand reported that mitochondria are bounded by a membrane and that they have a system of parallel, regularly spaced inner ridges which were named cristae.

It is now clear that there are, in fact, two membranes around the mitochondria of plants, protozoa, molluscs, insects and man: an outer membrane, set off by a fluidfilled gap; and an inner membrane which is folded inward at various points to increase its surface, forming the cristae, so that the enzyme molecules of the electron transport chain, which are attached to this inner surface in specific sequences, can be packed more densely side by side in the mitochondria.

This general design for mitochondria seems to have existed unchanged, through the evolution of plants and animals, for more than 1.2 billion years. Though mitochondria are generally visualized as ovalshaped, they can change shape quite drastically. They swell and contract in response to various hormones and drugs. Even in very low concentrations, Lehninger found, the thyroid hormone thyroxin will make mitochondria swell and will simultaneously "uncouple" the electron transport chain from the synthesis of ATP, so that even though carbohydrates may be metabolized and oxygen consumed, no ATP will be manufactured.

ATP, on the other hand, makes mitochondria contract. This swelling and contraction appear related to the movement of water in, out and through cells, which is of particular importance in the kidney. The more scientists learn about mitochondria, however, the more their curiosity is aroused, for these are without a doubt the strangest of the cell's organelles.

To begin with, mitochondria contain their own hereditary material (DNA), which resembles that of bacteria far more than that of animal cell nuclei. Because of this similarity, many scientists believe that mitochondria are derived from bacteria that infected a primitive cell and evolved into a useful symbiotic relationship with it. As Lewis Thomas puts it, "A good case can be made for our nonexistence as entities.

We are shared, rented, occupied. At the interior of our cells, driving them, providing the oxidative energy that sends us out for the improvement of each shining day, are the mitochondria and in a strict sense they are not ours. They turn out to be little separate creatures, the colonial posterity of migrant procaryotes, probably primitive bacteria that swam into ancestral precursors of our eucaryotic cells and stayed there.

Ever since, they have maintained themselves and their ways, replicating in their own fashion, privately, with their own DNA and RNA quite different from ours. They are as much symbionts as the rhizobial bacteria in the roots of beans. Without them, we would not move a muscle, drum a finger, think a thought. My mitochondria comprise a very large proportion of me. Looked at in this way, I could be taken for very large, motile colony of respiring bacteria, operating a complex system of nuclei, microtubules and neurons for the pleasure and sustenance of their families and running, at the moment, a typewriter, continues Thomas somewhat facetiously.I am obliged to do a great deal of essential work for my mitochondria. My nuclei code out the outer membranes of each and a good many of the enzymes attached to the cristae must be synthesized by me. Each of them, by all accounts, makes only enough of its own materials to get along on and the rest must come from me. And I am the one who has to do the worrying.

During the past few years, scientists have begun to learn which proteins and enzymes of the mitochondria are synthesized according to directions in the mitochondrial DNA and which are controlled by the genes in the nucleus. This information would be of considerable importance for the design of antibiotics and drugs, as well as for the understanding of genetic diseases. For example, researchers have found that chloramphenicol–a powerful antibiotic which can cause a fatal anemia–interferes with the synthesis of mitochondrial DNA, while other antibiotics do not. Mitochondria are just as important to the life of each cell as the heart is to the human body.

At least 95 percent of our cells' energy comes from mitochondria. (Only the mammalian red blood cells, which have no nuclei, get along without mitochondria; they obtain whatever energy they need from the partial breakdown of glucose in their cytoplasm). Growing cells continuously synthesize nucleic acids, proteins and other components, for which they need the energy that is stored in molecules of ATP.

Muscle cells which carry out mechanical work, plant cells which pump water against the force of gravity and many other cells also require a continuous source of energy. Besides supplying this energy, mitochondria apparently help to control the concentration of water, calcium and other ions in the cytoplasm.

They are also deeply involved in the breakdown and recycling of sugars, fatty acids and amino acids to recover their energy, as well as in the formation of urea as a waste product, to be excreted through the kidney. These strange organelles may thus hold clues to a variety of seemingly unrelated diseases. They have been linked to some liver diseases (*e.g.*, viral hepatitis, obstructive jaundice and cirrhosis of the liver); some muscle diseases (*e.g.*, familial myotrophic dystrophy); and some kidney diseases.

In all these conditions, pathologists have seen large amounts of materials pile up inside the mitochondria as "inclusions" of varying shapes and sizes, presumably waiting to be processed. What these inclusions consist of remains a mystery, but if scientists could find out, they might discover which specific components (for instance, enzymes) are involved in each disease. In addition, several calcium disorders may be related to mitochondrial defects. For example, abnormal concentrations of calcium (possibly caused by malfunctioning mitochondria) may lead to arthritis, bursitis, or arterial disease.

While all these leads are promising, the greatest spinoff from research on mitochondria may be in the area of energy–not only for the cell, but for general use by mankind. As the oil crisis has made us realise, we presently depend on a limited supply of fossil fuels–coal and oil which are of biological origin but cannot be replenished.

On the other hand, nature contains a far greater resource: a molecular know-how and technology gained over millennia, by which biological systems manufacture ATP from solar energy. This technology awaits us. It exists in very similar, if not identical forms in mitochondria and in their mirror image, the chloroplasts, which are found in green plants, where they do the work of photosynthesis, converting the sun's energy into carbohydrates.

Like mitochondria, chloroplasts have an outer and an inner membrane. They have some of their own DNA. They, too, may have originated as some kind of infection, probably with blue-green algae. And it is on chloroplasts that we ultimately depend for nearly all our energy–that of our bodies as well as that which is stored in deposits of oil and coal. If scientists ever unravel the details of the process by which chloroplasts store solar energy in carbohydrates and mitochondria release it, we may be able to produce unlimited quantities of usable energy.

MICROTUBULES, THE CELL'S PHYSICAL PROPS

All the organelles described so far are bound by membranes. For a long time, the rest of the cytoplasm–a liquid called the cell sap, or ground substance–appeared totally unstructured. But as scientists learned to use newer and gentler fixatives to prepare cells for electron miscroscopy, more and more tiny structures materialized in this soup. The most prominent of the new structures are two organelles which provide an intricate system of physical support for the cell– the equivalent of buttresses or skeletal bones– as well as a contractile mechanism for cell movement: microtubules and microfilaments.

Neither is bound by a membrane. Microtubules were first noticed in the middle 1950's, but they were seen only rarely until the development of glutaraldehyde as a gentle fixative in 1963. Extremely thin cylinders, about 200 to 300 angstroms in diameter and of variable length, microtubules are constructed chiefly of proteins called tubulins. In 1967, Edwin Taylor and his associates at the University of Chicago discovered that colchicine–a chemical used to arrest cell division–did its work by binding to the protein tubulin.

This pointed to the role of microtubules in mitosis and helped to answer a question which had worried generations of biologists: How do chromosomes sort themselves out into two sets and separate during cell division? The details of mitosis would have been very hard to explain without understanding the chemistry of the spindle fibres which organize this separation. In fact, if there were no microtubules, scientists would have had to invent them.Something of this sort was needed to account not only for mitosis, but for the surprisingly firm structure of such seemingly vulnerable fibres as the long, thin axons of nerve cells, as well as for the unique structure of some red blood cells, which are held in a disc-like shape by hoops composed of microtubules. Microtubules are also involved in assisting the transport of substances in and out of the cell and in the motion of cells themselves.

They make up the cilia and flagella, whip-like filaments which project from the surface of cells and move rhythmically. Large numbers of cilia are found on cells that line the respiratory system, for instance, where they help to sweep out dust and debris. Both cilia and flagella play an important role in human reproduction: the coordinated beating of cilia in the oviduct produces a sort of current which draws the female's egg into the uterus, while the rapidly thrashing tail of sperm actually is a flagellum.

TRANSPORT INTO AND OUT OF PLANT MITOCHONDRIA

Plant mitochondria must exchange metabolites and information (signal transduction) with the cytosol. To give some central examples, the end products of glycolysis, pyruvate and malate, need to be taken up, and citric acid cycle intermediates exported to be used as building blocks in biosynthetic processes. Likewise, ADP and P_i must be taken up and ATP exported for the mitochondrion to fulfill its role as the powerhouse of the cell.

To carry out the transport of this large number of molecules, plant mitochondria possess a family of mitochondrial carrier proteins. These are transmembrane proteins with a common general structure and they typically transport relatively small charged molecules, like metabolites. In Arabidopsis there are almost 60 members, most of which are not functionally characterized, giving a hint to the complexity of the mitochondrial transport processes. However, mitochondria exchange more than metabolites, and therefore also other types of transporters are involved.

The outer membrane is permeable to molecules of less than 10 kDa, (i.e., all non-polymeric compounds). The permeability results from the presence of pores, which appear to be open in most *in vitro* experiments. Recent studies in animals have characterized pore-forming proteins in the mitochondrial outer membrane, called porins, which appear to regulate transport across the membrane.

Plant mitochondria also have porins but their characterization awaits further research. In our discussion here we will focus on the regulation of mitochondrial transport across the inner mitochondrial membrane. We will first consider the relatively few molecules that can move across the inner membrane without the help of a specific carrier protein. We will then discuss the various metabolite transporters and, finally, briefly look at the transport of proteins and RNA.

Small, Neutral, and Hydrophobic Molecules Diffuse across the Inner Membrane

Irrespective of size, molecules that either are charged or are very hydrophilic can only cross a lipid bilayer with the help of a specific protein carrier.

Small, *neutral* molecules can diffuse across the lipid bilayer and do not require protein transporters. Among those are O_2 and CO_2, two important molecules for respiration. O_2 is needed on the inner surface of the inner membrane where the active sites of both cytochrome *c* oxidase and the alternative oxidase are located. The CO_2, produced as an end product in the citric acid cycle, needs to diffuse out of the mitochondrion, ultimately to be lost to the surroundings or re-fixed in the chloroplasts. In addition, however, it is possible that CO_2 can also be transported out of the mitochondrion as bicarbonate (HCO_3^-). Mitochondria are osmotically active and swell in a hypotonic medium. This means that the inner membrane is also permeable to H_2O, but it is not clear whether this occurs by diffusion or if aquaporins may be involved.

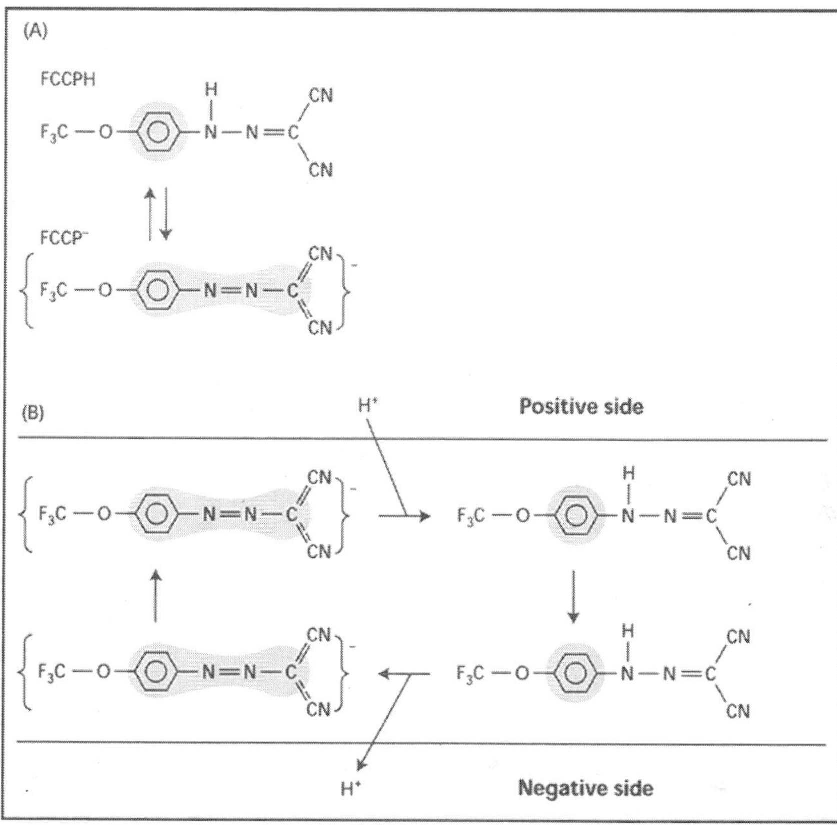

Fig. The Mechanism of action of Uncouplers (also known as protonophores).

(A) The protonation/deprotonation of FCCP (trifluoromethoxy carbonyl cyanide phenylhydrazone), which is a weak acid. (B) In the presence of an electrochemical proton gradient across the membrane, FCCP will become protonated and thus pick up a proton on the positive side of the membrane, move across the membrane in the neutral form, and lose the proton on the negative side. This is driven by the proton gradient. In the negative form, FCCP will be driven back out across the membrane by the membrane potential. The net result is the movement of one proton from the positive to the negative side of the membrane, which results in the dissipation of the proton gradient, and the movement of one electrical charge from the negative to the positive side of the membrane, which results in the dissipation of the electrical gradient. The shaded area represents the system of conjugated double bonds (p-orbital system).

For technical reasons, it is of importance that several well-known inhibitors of mitochondrial function can diffuse into mitochondria because they are uncharged and hydrophobic. These are rotenone, the specific inhibitor of complex I, antimycin, the specific inhibitor of complex III, HCN, the protonated form of the cyanide (CN) inhibitor of complex IV (and other heme-containing enzymes), and oligomycin, the specific inhibitor of the F_0F_1-ATP synthase (complex V).

Chemical uncouplers are very valuable to study membrane bioenergetics because their structure with conjugated double bonds allow them to diffuse across the membrane both in the protonated form and the unprotonated form. A case in point is FCCP (trifluoromethoxy carbonyl cyanide phenylhydrazone), a very efficient mitochondrial uncoupler, as illustrated in Above Figure.

Transport of Inorganic Ions

Inorganic phosphate (H_2PO_4) is taken up into the matrix in exchange for a hydroxyl ion (HO). The energy for the uptake is supplied by the proton gradient ("pH). We will see below that the phosphate gradient, in turn, drives transport of other metabolites.

Several other inorganic ions—Mg' Ca, K, and NH_4 —are involved in mitochondrial metabolism, but the mechanism by which they pass the inner membrane of plant mitochondria is not well understood. In mammalian mitochondria, uptake of K through a K -specific channel, and extrusion via a K /H antiporter, is involved in matrix volume regulation and cellular signaling. A similar system of K transport is present in plant mitochondria.

Both mammalian and plant mitochondria contain an uncoupling protein (UCP), as discussed briefly in Topic. This protein facilitates the movement of protons across the inner membrane and therefore uncouples electron transport and decreases ATP yield.

Transport of Adenylates

One of the important functions of mitochondria is to supply the cell with ATP. To do this, mitochondria have an adenine nucleotide transporter that facilitates the exchange of ADP (inward) for ATP (outward).

This exchange is driven by the membrane potential (inside negative), since there is a net outward movement of one negative charge. In combination with the phosphate carrier reaction, this means that the uptake of one ADP and one phosphate and the export of one ATP (i.e., the transport event connected to mitochondrial ATP synthesis) costs one vectorial proton—the electrical component for adenylate exchange and the "pH component for phosphate uptake.

All of these transporters are directly or indirectly driven by the electrochemical proton gradient. P_i uptake is driven by the proton gradient (hydroxyl anion in the opposite direction) and the resulting P_i gradient is used to drive the uptake of dicarboxylate anions on the dicarboxylate transporter. The dicarboxylates can, in turn, exchange for 2-oxoglutarate or citrate. In the latter case, citrate is protonated on one of its acid groups to maintain electroneutrality. The exchange of glutamate (in) for aspartate (out) is electrogenic in mammals and driven by the electrochemical proton gradient since glutamate is transported as a monovalent anion against aspartate as a divalent anion. The latter transporter is important for the operation of the malate/aspartate shuttle.

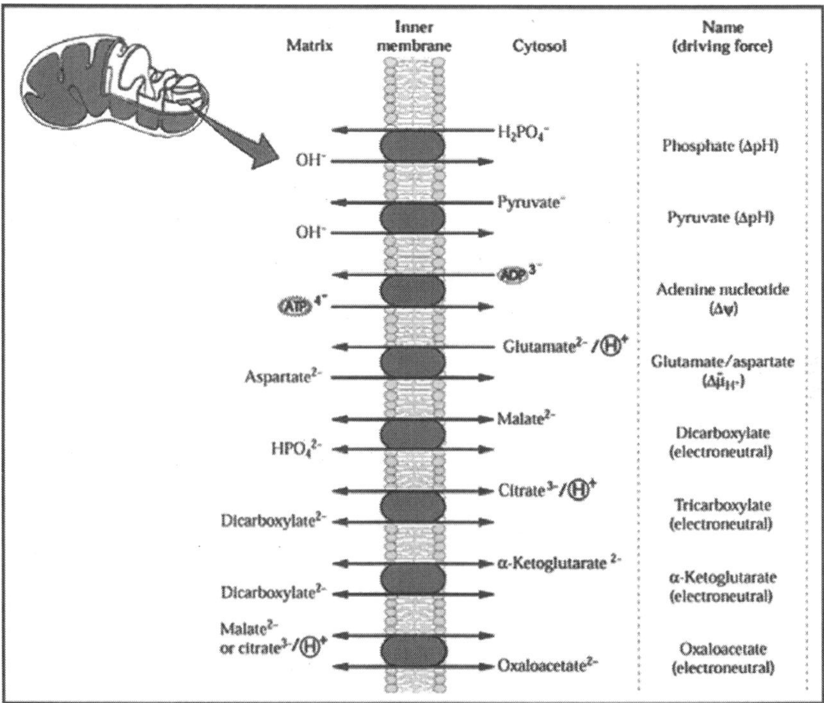

Fig. Transporters in the Inner Membrane of Plant Mitochondria.

Transport of Citric Acid Cycle Intermediates and Related Compounds

Pyruvate and malate, the end products of glycolysis, both have transporters in the inner membrane. Pyruvate is taken up in exchange for a hydroxyl ion, and malate in exchange for H_2PO_4. The transporter for malate also takes succinate and is therefore called the dicarboxylate transporter.

Malate or succinate can, in turn, exchange for citrate on the tricarboxylate transporter. Malate can also exchange for 2-oxoglutarate via a special transporter. An interesting transporter, the oxaloacetate transporter, exchanges oxaloacetate for malate (or citrate), which permits the shuttling of reducing equivalents.

The joint activity of these transporters allows plant mitochondria to import or export most citric acid intermediates. Finally, the glutamate/ aspartate transporter facilitates the uptake of glutamate. In mammalian mitochondria, glutamate is taken up together with a proton, and it is thus driven by both the electrical and the chemical proton gradient. A similar transporter is present in plant mitochondria, but a proton co-transport has not been shown, so the transport may be passive in plants. Recently, several additional transporters for metabolites have been identified, and it is likely that certain setups of transporters are only expressed under special developmental conditions, like the mobilization of storage nutrients in developing seedlings.

In several cases the involvement of the mitochondrion in cellular metabolism implies that certain metabolites are taken up across the inner membrane, but the transporter has not yet been described. This is the case for the massive flux of glycine into—and CO_2, serine and NH_4 out of—mitochondria in C_3 leaves during photorespiration. This is also the case for proline, which accumulates to high concentrations in the cytosol during episodes of plant water stress, and it is then broken down during the rehydration phase by a pathway involving two mitochondrial enzymes.

Transport of Coenzymes

Coenzymes are either synthesized inside the mitochondrion or imported. There is some evidence that, in addition to ATP, at least NAD , thiamine pyrophosphate, and coenzyme A are imported across the inner membrane.

The final step in ascorbic acid biosynthesis has recently been shown to be on the outer surface of the inner mitochondrial membrane. In view of the accumulating evidence that ascorbate has an important function in removing ROS inside the mitochondrion, we should expect to find an ascorbate transporter in the inner mitochondrial membrane.

The situation with folate (involved in 1C transfers) is the opposite to that with ascorbate: The only site of synthesis of many of the folate metabolites in the plant cell is the mitochondrial matrix. We can therefore expect that these molecules are exported via specific transporters.

Transport of Reducing Equivalents—Metabolic Shuttles

The two shuttles shown have both been demonstrated to work in isolated mitochondria. (*Top*) The malate/aspartate shuttle uses isoforms of two enzymes, malate dehydrogenase (1), and aspartate aminotransferase (2), to interconvert malate and aspartate, which are then exchanged using two transporters, the glutamate/aspartate (A) and the 2-oxoglutarate (B). The electrogenic glutamate/aspartate transporter provides directionality so that reducing equivalents are transferred into the mitochondrial matrix. (*Bottom*) The malate/OAA shuttle uses malate dehydrogenase isoforms (1) in the matrix and the cytosol to interconvert malate and OAA, which are then exchanged by the OAA transporter (C). This shuttle can transfer reducing equivalents in either direction depending on the requirements of the cell. Mammalian mitochondria cannot oxidize cytosolic NADH or NAD(P)H directly, so they make use of "shuttle mechanisms." An exchange of metabolites by inner membrane transporters takes place through these shuttles so that a reduced compound moves into the matrix in exchange for a more oxidized compound moving out into the cytosol. In this way, reducing equivalents are transported into the matrix ultimately to be oxidized by the respiratory chain. The shuttles can also be used to transport reducing equivalents in the opposite direction.

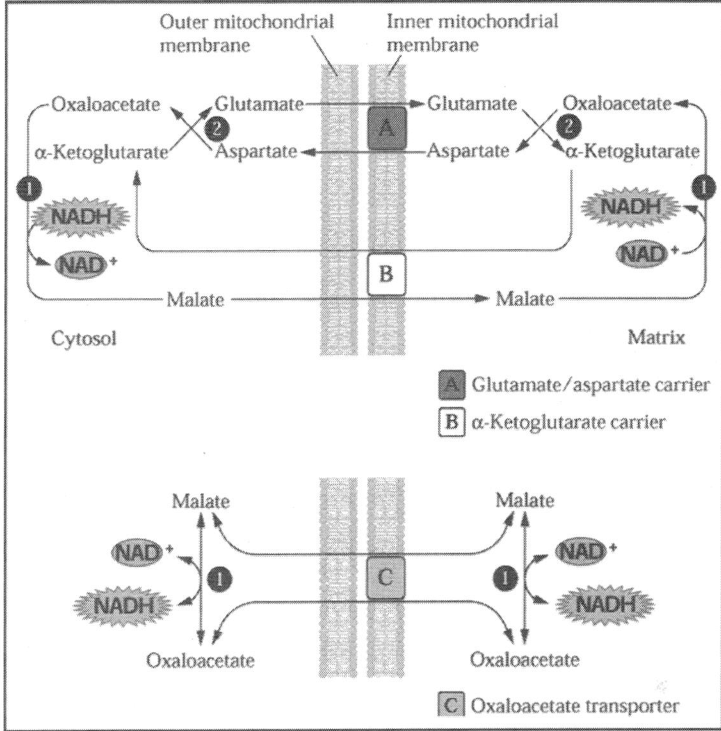

Fig. Metabolite shuttles across the inner membrane of plant mitochondria.

Although plant mitochondria can oxidize cytosolic NADH and NAD(P)H directly by the two external NAD(P)H dehydrogenases, they can also use these two metabolic shuttles. The malate/oxaloacetate shuttle uses malate dehydrogenase in the cytosol and in the matrix to catalyze the interconversion of malate (reduced) and oxaloacetate (oxidized). These two compounds are exchanged on the oxaloacetate transporter.

This exchange, however, is not driven by the electrochemical proton gradient across the inner membrane, so it can only move reducing equivalents from a relatively reduced compartment to a relatively oxidized compartment. Since the mitochondrial matrix is more reduced than the cytosol, the oxaloacetate transporter is likely to work primarily in the export of reducing equivalents.

Under photorespiratory conditions, large amounts of glycine are oxidized in the mitochondria to give NADH while, at the same time, NADH is required in the peroxisomes to reduce hydroxypyruvate to glycerate. It is thought that as much as 50% of the matrix NADH is exported, probably via the malate/oxaloacetate shuttle.

The more complex malate/aspartate shuttle involves two matrix enzymes and two cytosolic enzymes as well as two inner membrane transporters. In mammals, it is powered by the electrochemical proton gradient in the direction of the import of reducing equivalents. However, since the glutamate/aspartate exchanger may be passive in plants, this shuttle in plants may similarly be independent of the electrochemical proton gradient. If so, it would only be able to function as an exporter of mitochondrial reductant.

Transport of Proteins—Protein Import

Plant mitochondria contain their own DNA (mtDNA) and ribosomes and are capable of synthesizing the proteins encoded in the mtDNA. However, plant mtDNA encodes only a few of the proteins found in the mitochondrion. The great majority of the mitochondrial proteins have to be imported from the cytosol where they are synthesized on free ribosomes. These are synthesized with a presequence or targeting peptide in the N-terminal end, which is recognized by receptors on the outer mitochondrial membrane.

The import of proteins requires a complex machinery with two multisubunit protein complexes—one in the outer membrane (TOM, transporter of the outer membrane) and one in the inner membrane (TIM, transporter of the inner membrane). Concomitant with, or immediately afterwards the import of the protein into the correct mitochondrial compartment, the targeting peptide is often cleaved off to yield the mature and active protein.

Transport of tRNA

Mammalian mtDNA encodes the full complement of tRNA required for mitochondrial protein synthesis. In contrast, plant mtDNA only encodes some of the tRNAs required for protein synthesis within the mitochondrion. The missing tRNA must be imported by a yet-to-be-elucidated import mechanism.

EXPANSION OF PLANTS CELL

Within 5-10 minutes of application of auxins to isolated epicotyls, or intact tip of the plant or excised segments, growth in their length starts. This early effect is rapid and spontaneous. This rapid growth continues up to 30-45 minutes and then slows down but remains steady. This is also accompanied by an increase in cellular respiration, protein synthesis and the rapid extrusion of protons or in particular Hydrogen ions through the cell walls.

Another significant observation of this early rapid growth is insensitive to the inhibition of transcription (RNA synthesis) and translation (protein synthesis) - ex. Actinomycin-D at high concentration inhibits RNA synthesis completely and cycloheximide (CHI) inhibits cytosol protein synthesis. This clearly suggests that the early rapid growth requires neither the synthesis of RNA nor the synthesis of proteins. But this early effect of auxin is totally inhibited by colchicines at 0.5 to 1 per cent concentration. Colchicine, an alkaloid from the tubers of Colchicum autumnale brings about the inhibition of tubulin (TB) polymerisation into long microtubules, which act as the cytoskeleton for mechanical support in the cytoplasm. The colchicines effect on early growth further indicates the important role played by the assembly of microtubules in providing a mechanical structure and the force for all expansion.

Increased metabolic activity, particularly the degradation of osmotically inactive starch, etc., leads to the increase of osmotic pressure (they are related to each other) and a steep DPD gradient is created. This facilitates the inward diffusion of water; consequently turgour pressure builds up within the cells. This force acts on all sides of the cell wall and forces the cell to expand in length and breadth. However recent investigations have shown that the cell elongation takes place under negative pressure.

The above features suggest that the process or the mechanism of cell elongation is complex and multiphased, but highly regulated. Various theories have been proposed from time to time and none of them could prove their claim because of one or the other drawbacks. Here we will restrict to a comprehensive account of cell elongation.

Early phase: The auxin's effect in the early phase on cell growth is rapid. This is insensitive to the inhibitors of transcription and translation but insensitive to colchicines and respiratory inhibitors like cyanide, DNP etc. From this it can be deduced that cell elongation does not require new RNA and protein synthesis, but it requires ATP and ATP dependent tubulin polymerisation into microtubular cytoskeleton. Auxin at the early stages activates and increases the rate of respiratory metabolism.

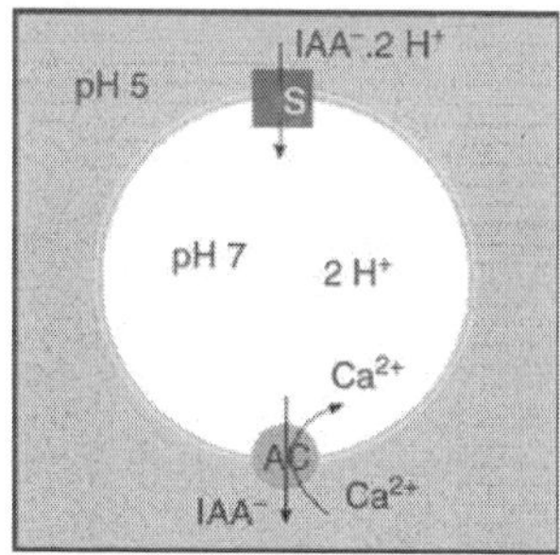

With the result ATP pool builds up rapidly. Auxin also activities membrane bound nucleating centre of tubulin polymerisation. These nucleating centres, utilising ATP and tubulin monomers found in the cytoplasm, start polymerising and soon build up long microtubules.

These, in turn, start orienting parallel to the long axis of the cell, which probably requires cytochalasin-B sensitive microfilaments. With the rapid increase in microtubules, oriented in longitudinal plane, considerable force is exerted on the cell boundaries.

As the microtubular polymerisation is building up, ATP ase/H* gets activated, and H* ions are pumped out into the interspaces of cellulose fibrils acidic pH in turn activates cellulases enzymes, which starts loosening up the cell-wall cellulose fibres. This loosening up of cell wall concomitant with acting as a mechanical force-the cell expands in length. This process starts 4-5 minutes after application of auxins and the elongation continues rapidly for about 20-30 minutes.

The late phase starts around 30-40 minutes after auxin application. This is slower than the first phase, but growth is steady. This phase is sensitive to Actinomycin-D, CHI, Colchicine and respiratory inhibitors, indicting that this process requires metabolic energy, tubulin-polymerisation, RNA synthesis as well as protein synthesis. As tubulin pool gets exhausted in the first phase, it has to be replaced. That means it requires the synthesis of RNA and proteins.

Apart from these proteins, the cytoplasm requires the synthesis of other proteins, for the increase in the mass of protoplasm. It is quite possible that the early activation of early translation may have significant effect on gene activation. Thus it takes some time for the second phase of growth to set in. The second phase provides all materials for sustained growth.

Growth Movements: The phenomenon of bending movement of plants in response to light stimulus is called Phototropism, the growth of roots towards soil is called Geotropism.

The curvature growth due to the touch is named as Thigmotropism, etc. Such differential growth movements are due to differential concentration of auxin. The stem tips and root tips have different optima of their auxin requirement to show maximum growth.

The mechanism of these movements will be explained in the chapter on growth movements. Auxin brings about phototropic movement because differential accumulation of auxin at exposed and darker side. *Root initiation*: Vegetative propagation of stem cuttings is a normal practice in horticulture, floriculture and agriculture. Out of some stem cuttings which are put into soil, a few produce roots easily but others may or not produce roots. The roots that develop from any part of the plant body other than the radicle are called adventitious roots.

The development of such roots from the stem or any other part is a phenomenon of dedifferentiation. Stem is quite different fro the root with respect to its structure and function.

Still some of the tissues found in stem undergo dedifferentiation and develop into roots. This dedifferentiation and growth of roots from the stem is initiated by auxins Indole acetic acid, IBA, NAA, and other hormones. When applied to stem cuttings they induce the formation of roots. IBA has been found to be very good in this regard.This indicates that the pre-existing m RNA protein complex is drawn towards protein synthesis: it shows increased rate of protein synthesis. Some of the proteins thus synthesised in turn activate differential nuclear gene activation. As a result of it, a cascade of events occurs. Generally there is an increase in the synthesis of all cellular proteins. At the same time there is an enhanced rate of synthesis of some specific proteins. Also some new proteins are synthesised. Among these, tubulin, a membrane bound fraction, is synthesised at a high rate. It is believed that this protein along with microfilaments gets assembled into microtubules which orient themselves in a specific way so as to facilitate the polarity fixation for cell division.

In response to auxins, some of the cells in the pericycle region get stimulated and the required macromolecules are synthesised by differential gene activation. Thus these cells endowed with these new potentialities divide and redivide, and soon they organise into root primordial. They in turn develop into new roots which penetrate through cortical cells. The time taken for this new root formation, in response to auxin supply is just 24-36 hours. This effect of auxin induced new root formation has been well exploited by horticulturists and others. The following diagram shows how auxins interact with cytokinins in eliciting growth responses. *Apical dominance*: The active growth of terminal bud inhibits the development of lower axillary bunds. This effect is called apical dominance. Many plants like conifers and other exhibit the apical dominance so predominantly that the plant grows tall with few branches and appear conical in shape. But plants like banyan, mango, tamarind do not show that much of apical dominance, hence these plants spread outwardly. This phenomenon of apical dominance has been attributed to the differential concentration of auxin and nutrient supply to the axillary buds.

Abscission: With the onset of winter season the leaves of deciduous plants turn yellow and fall off. The falling of the leaves is due to the formation of abscission layers at the base of the petioles. This layer consists of thin walled but suberised cells. The middle wall gets disintegrated and the leaves fall down. The development of this layer is stimulated by a plant hormone called ABA. But the application of auxin to such leaves prevent the formation of abscission layer and thus prevent the falling off the leaves. This has a greater application particularly in preventing the premature falling of fruits like apple, orange, etc.

Parthenocarpy: Development of fruits without fertilization is referred to as parthenocarpy and such fruits are seedless. Normally plants do not produce parthenocarpic fruits. Normally plants do not produce parthenocarpic fruits.

But the application of auxin to flowers induces parthenocarpic fruit setting and prevents premature falling of fruits as well. Plants with triploid and many polyploid genomes do not set fruits. If such plants are sprayed with auxins at the time of flowering, fruit setting is enhanced. The fruits thus developed are large, seedless, and sweeter. Seedless oranges, lime, watermelon, sapota, ramphal and seethaphal, if they are produced this way, they have good market at the commercial level.

Gibberllic Acid

Gibberllic Acid is another natural hormone found distributed in higher plants. This was first discovered in Japan, later its structural and functional details have been elucidated by the people in the west. So far 52 or more different kinds have been identified. Among them GA2 2GA1 are the most common ones found in all plants.

GA is a steroid like molecule in its structure. Young leaves are the sites in which this hormone is mediated by phytochromes and then it is transported to the regions of growth.

The most significant effect of GA on the growth of plants is it effects internodal cell elongation profoundly than any other hormones. Even genetic dwarfs (eg. Dwarf Pisum sativum) grow as tall as the natural tall plants.

It also brings about a dramatic effect called Bolting and Flowering. Long day plants (which require 12-14 hrs of day light treatment for flowering) under short day conditions produce a cluster of leaves on highly shortened stem and never produce flowers. When such plants are sprayed with GA, the stem starts elongation with enhanced growth of internodes. Simultaneously the terminal buds starts producing floral branches. This phenomenon is called bolting and flowering. However both bolting and flowering are separated by time and space. Besides these above said effects, GA is essential for the mobilisation of food reserves for the germinating cereal grains like wheat and maize. At the time of germination, the young embryonic axis synthesises this hormone, which is soon released into endosperm and reaches the aleurone layer.

It is this layer, GA activates differential gene expression. As a result, specific m.RNAs is transcribed and later specific proteins like amylase are synthesised on translation. The amylase thus synthesised, acts upon the starch found in the endosperm, and the glucose thus released is drawn into the embryonic tissue which is in dire need of energy source. Notwithstanding the above effects, GA also effects parthenocarpy, rooting, germination, etc. This hormone also interacts with other natural hormones like auxin and cytokinin and brings about differential growth.

CYTOKININS

Another class of growth regulating hormones has been discovered by Miller et al and they are called Cytokinins. They are the derivatives or nucleotides. 6-Furfuryl amino purines, Benzyl amino purines and such compounds act as cytokinins.

Cytokinins are generally synthesised in the root tips, and then they are translocated to other regions. The endosperm of coconut milk, Borassus and of other plants is a rich source of cytokinins. Cytokinin level in plants is further regulated by other hormones like Auxins and GAs. On the contrary cytokinins control the synthesis of other hormones like Abscisic acid, a growth inhibitor.

Cytokinin *per se* has an important role in the cytokinesis. But along with auxins, it effects differentiation of plant organs from the callus. Cytokinin by itself inhibits root formation, but in the presence of auxins, it induces organ formation like roots and shoots. If the ratio between auxin and cytokinin is high, roots are differentiated, but the lower ratio induces the formation of shoots.

The intermediate ratio stimulates the development of these hormones in differentiation at the molecular level is very interesting by unfortunately it is least understood. Cytokinins are also found to break dormancy of winter buds as well seeds. The exact mechanism by means of which they over come the dormancy is not known, still it has been thought, that they effect through the suppression of IBA which is responsible for dormancy. Another interesting effect that the cytokinins produce is the prolongation of senescence (Ageing). This effect is often referred to as Richmond and Lang's effect.

7

Cellular Metabolism

It is generally recognized that the energy necessary for cellular activities, whether it be used to maintain the steady state of adult life, the "dynamic equilibrium" of Schoenheimer, or the state of growth and synthesis which predominates in dividing and growing cells, is provided by enzymatic reactions largely of the oxidation-reduction type.

This has been amply demonstrated by the increased O_2 utilization during growth and by the necessity of aerobic processes for the synthesis of carbohydrates, fats, and proteins which together make the bulk of the cells. It will be therefore necessary before discussing the role of enzyme reactions in cellular growth to review some fundamental aspects of the properties of the oxidation-reduction enzymes.

GENERAL PROPERTIES OF ENZYMES

Enzymatic oxidation-reduction reactions are distinguished from chemical reactions in general by the facts that they are reversible, that electron exchange occurs in steps, and that they have a great tendency to form coupled reactions.

The reversibility of enzyme reactions can be accepted as a general property, since more and more of those reactions hitherto considered irreversible are being shown to be reversible.

Witness the enzymatic synthesis of phosphopyruvic acid from pyruvic acid, a reversal of the dephosphorylation of phosphopyruvic acid, which had been reported as an irreversible process.

In many instances reversibility of oxidation-reduction reactions is made easy by the entrance of P into the reactions and generation of energy-rich phosphate bonds, as was shown by Warburg and Christian in the compulsory coupling between the oxidation of phosphoglyceraldehyde and the phosphorylation of adenylic acid, a finding extensively generalized to other reactions by Lipmann.

The first step in an enzymatic reaction is the formation of the enzymesubstrate complex. Evidence in favour of such a concept, introduced by Michaelis and Menten, is abundant. Knowledge of the nature of this complex formation awaits clarification of the structure of proteins, since it is probably the protein moiety which combines with the substrate. The substrate to be oxidized must fit quite closely to the side chains of the protein molecule where combination occurs. No lock, however, remains immune to the prying of false keys; and substances with chemical structure close to that of the substrate will compete for union with the protein, thus affecting the activity of the enzyme.

This is strikingly demonstrated by the inhibition of acetate oxidation by fluoroacetate discovered in animal tissues and in yeast. In yeast, acetate oxidation, completely inhibited by monofluoroacetate, was not affected at all by the other halogen acids, bromo-, iodo-, chloro-acetic acids, as well as by trifluoroacetic acid.

Table. Effect of Halogen Acetates on the Oxidation of Acetate by Baker's Yeast. pH, 6.2. Acetate Concentration, 0.01 M; halogen acetate, 0.001

	Inhibition (%)		
Halogen	*Acetate Oxidation*	*Endogenous Respiration*	*Interatomic Distances C -- Halogen bond (A)*
CH2FCOOH	95	24	1.41
CH2ClCOOH	none	none	1.76
CH2BrCOOH	4.8	40	1.91
CH2ICOOH	1.5	none	2.10
CH3COOH	—	—	1.09
CF3COOH	none	none	—

This inhibition is due to competition of fluoroacetate with acetate in the protein of the enzyme, for inhibition occurs in the first step of acetate metabolism, namely citric acid formation. Of all these acids it is monofluoroacetic acid which approaches most closely in size acetic acid, as the interatomic distance between C and F (1.41 A) is only 0.32 A more than the interatomic distance between C and H.

An increase in the interatomic distance of 0.35 A as in C and Cl (1.76 A) was enough to avoid penetration of the halogen acid into the space reserved for acetic acid in the protein moiety. The protein-substrate complex formation resembles thus strikingly the antibody-antigen complex formation in the specificity.

Some proteins require the presence of —SH groups for enzymatic activity. Furthermore, in some cases only a portion of the —SH groups present in the protein are necessary for enzyme activity, as demonstrated with urease by Hellerman and coworkers.

Yeast cells, 8.5 mgs.; Mg acetate, 0.077 *M*; fluoroacetate, 0.005 *M*. Duration of experiments, 4 hours. Blank values for oxygen uptake and citric acid have been subtracted.

Table. Effect of Fluoroacetate on Citrate Formation by Baker's Yeast.

Experiment No.	*Citric acid formation (cmm.)*		*Inhibition (%)*	*O2 Uptake (cmm.) Control*	
	Control	*Inhibitor*		*Control*	*Inhibitor*
I	4.99	0.045	91	1010	0
II	3.86	0.038	90	1005	0
III	5.0	1.4	72	1012	242

A survey made by Barton and Singer showed that a large number of enzymes which catalyze the metabolism of carbohydrates, fats and protein derivatives need the electronegative—SH groups for activity. Doubtless there are still other enzymes, not yet studied, which belong to this class. An indication in favour of this assumption is given by the inhibition of mitosis produced by small doses of x-rays and by nitrogen mustards. Up to the time of Dale's experiments little attention was paid to enzyme inhibition by x-rays. In 1940 Dale demonstrated that enzyme inhibition could be produced by small doses of x-rays provided protein concentration was kept low.

Barron, Dickman, and Singer have shown that x-rays, α, and β particles and γ radiations inhibit sulfhydryl enzymes at small doses by oxidation of the —SH groups of the protein. Inhibitions produced by x-rays and α particles could be released on addition of glutathione. This inhibition is undoubtedly produced by the OH and OH_2 radicals and H_2O_2 formed by the ionization of water during irradiation.

In fact, protection of the —SH groups by mercaptide formation with Cl-Hg benzoic acid abolished inhibition of urease by γ radiation. Destruction of H_2O_2 by catalase, diminished the degree of inhibition of phospho-glvceraldehyde dehydrogenase by a particles, and suppressed the inhibition by β particles. The combination of nitrogen mustards with —SH groups is known.

Table. Effect of X-Rays on the Activity of Phosphoglyceraldehyde Dehydrogenase. Buffer, phosphate, pH 7.0. Enzyme activity, where t, time in minutes; Co (DPN) initial; C (DPN) at time t.

X-Ray dose (r)	*Enzyme Activity K × 105*	*Inhibition (%)*	*Reactivation with Glutathione (0.002 M)*
none	4.80	—	—
25	4.80	none	—
50	4.80	none	—
100	3.81	21	complete
200	2.43	50	62
300	0.98	80	—
400	0.98	80	—
500	0.25	94	10

However, as this combination is irreversible, it is difficult to establish definitely whether enzyme inhibition is due to this reaction or to combination with some other group of the protein. Very little is known about the role of these—SH groups.

Protection of the—SH groups of the protein moiety of succinoxidase by succinate and by malonate would suggest that the substrate is anchored in close vicinity to the—SH groups.

Table. Inhibition of Crystalline Urease by Radium Radiation Effect of Glutathione, Glycine, and p-Cl-Hg Benzoic Acid.

Dosage (r)	*Additions*	*Inhibitions (%)*
100	—	24
100	glutathione	30
200	—	36
100	glycine, $1\times10^{-6}M$	none
100	glycine, $1\times10^{-1}M$	none
200	—	34
200	glutathione	39
200	p-C1-Hg benzoic acid+glutathione	none

The proximity of the carboxyl electronegative groups would decrease the activity of the —SH groups and thus render them less susceptible to the action of sulfhydryl reagents. Some other proteins require the presence of a metal (Mg, Mn, Zn, Co) for enzyme activity; the metal does not participate in the reaction.

Their role is still more obscure than that of the—SH groups. It is known, however, that—SH groups and metals contribute to increase the stability of the native protein.

The combination of the substrate with the protein brings forth the "activation" of the substrate. It is not yet clear what is meant by activation of the substrate. It is plausible, however, to postulate with Michaelis that activation consists in the formation of a radical, the first step in the process of oxidation of organic substances with a loss of two electrons.

This is implicit in Michaelis' theory of compulsory univalent oxidation. In condensation reactions where two substrates take part in the reaction, as in the formation of acetylcholine, it seems that only one of the substrates combines with the protein and becomes activated.

Evidence for this assumption is provided by inhibitions with competitive inhibitors. For example, methyl bis (dichloroethyl) amine HCI inhibits choline oxidase because of the similarity of its first transformation product, ethylene imonium, to choline; it also inhibits the synthesis of acetylcholine and its hydrolysis. It has no effect on the hydrolysis of tributyrin.

Fluoroacetate inhibits the oxidation of acetate while it has no effect on the synthesis of acetylcholine and acetylations in general. It must be concluded that in acetylations the protein combines with the acetylated compound and not with the acetyl group.

Table. Effect of Methyl his (Dichloroethyl) Amine (0.001 M) and Fluoroacetate (0.01 M) on Acetylations, Oxidations, and Hydrolysis.

Reaction	*Inhibition (%)*	
	N Mustard	*Fluoroacetate*
Acetylation of sulfanylamide	none	none
Acetylation of choline	75	none
Oxidation of choline	complete	none
Oxidation of acetate	none	complete
Choline esterase (hydrolysis of acetylcholine)	complete	--
Esterase (hydrolysis of tributyrine)	none	--

A number of proteins, components of enzyme systems, have been isolated and prepared in crystalline form. Furthermore our knowledge of the physical and chemical properties of proteins has considerably increased during the last years. The time has come to use these techniques for a serious study of the protein moiety of the enzyme.

It is necessary to know the architecture of the protein moiety of the enzyme, the spacing of the side groups to which the substrates attach themselves, and the nature of the activation of the substrate.

CELLULAR TOTIPOTENCY

Totipotency is the ability of all living cells potentially to regenerate whole new individuals - the trick is how to trigger them to do it and so to demonstrate it. If I remember correctly, Dolly the sheep was cloned from a cell from another sheep's udder - a differentiated cell, which in normal circumstances would have remained as such. Similarly in plants, it is possible to trigger single epidermal cells in oilseed rape, sunflower, and I am sure there are others, to develop into embryos and subsequently to whole plants. Even immature pollen grains have this potential - these immature cells can be switched to develop into whole plants, even though they only have one set of chromosomes, and give rise to haploid plants, although sometimes these may diploidise during culture, to generate homozygous diploid plants.

One of the hardest things in plant tissue culture is to get tissues from mature trees to illustrate totipotency, but by forcing a rejuvenation step, it is even possible to do this with some species - I believe quite a bit of work has been carried out on pines.

Simple cuttings, and the SAPS tissue culture system depends on this ability of cells to switch their development - hence root formation from hypocotyls, and root formation from woody cuttings, which allow new plants to be obtained for propagation. We described the different media used for cell/tissue culture, and also discussed culture of individual differentiated cells. However, in plant tissue culture experiments, unless single cell cultures are absolutely necessary, more often we use explants (pieces of differentiated tissues) to initiate their growth in culture.

This is also necessary to obtain callus from which single cell cultures are derived. During the last few decades, even for development of new crops or for improvement in the characteristics of existing crops (*e.g.* to obtain disease free crop, etc.), the techniques of tissue culture in combination with genetic manipulation have become important tools.

For instance, as an alternative to vegetative propagation, shoot tip propagation was used commercially for rapid and consistent reproduction of elite or difficult to propagate genotypes. , When an explant from differentiated tissue is used for culture on a nutrient medium, the non-dividing, quiescent cells first undergo certain changes to achieve a meristematic state. The phenomenon of the reversion of mature cells to the meristematic state leading to the formation of callus is called dedifferentiation. The component cells of callus have the ability to form a whole plant, a phenomenon described as redifferentiation.

These two phenomenons of dedifferentiation and redifferentiation are inherent in the capacity described as cellular totipotency, a property found only in plant cells and not in animal cells.In other words, while a differentiated plant cell retains its capacity to give rise to a whole plant, an animal cell loses this capacity of regeneration after differentiation. Although, generally a callus phase is involved before the cells can undergo redifferentiation leading to regeneration of a whole plant, but rarely the dedifferentiated cells can give rise to whole plants directly without an intermediate callus phase. These aspects of tissue culture and its application in clonal propagation and in the general on of hereditary variation called somaclonal variation.

Some plants have cells that are able to convert from a structured, or differentiated, state to a totipotent state. One example is the ZZ plant (Zamioculcas zamiifolia). This plant is so cellularly precocious that a mere piece of leaf or stem can be used to grow a new plant. Other plants and trees can produce spontaneous buds on the stem or trunk. These emerge from the activity of the meristematic tissue of the cambium layer, and since they are not buds that are normally produced in the axils of leaves, they are called adventitious buds. This can happen even when the tree in question is not the kind that will root from cuttings. The formation of adventitious buds begins with the production of a callus, which is essentially an undifferentiated lump of cells resembling a small tumor. Through genetic and hormonal processes, cells differentiate into growing points which then produce new stems with leaves. A similar process can result in adventitious roots as well.

Advances in the understanding of totipotency and how cells can be stimulated to become totipotent has resulted in an entire industry, the business of micropropagation, or what is commonly referred to as "tissue culture". While micropropagation labs do culture tissue (meristematic tissue), this work is not the same as what is done in a scientific tissue culture labouratory. The work done in scientific labs is much more esoteric and amazing. I studied the science of tissue culture when I attended graduate school, and we did things like isolating single cells from a callus and removing the cell wall using enzymes to yield a naked plant cell called a protoplast. These are important in scientific research because genes are more easily inserted into plant cells when they don't have a cell wall around them. We grew pure callus cultures and cell lines that were, essentially, all totipotent cells. Using the right combinations of plant hormones, we could then cause the cells to turn into plant shoots with roots. These and other experiments done in scientific tissue culture labs are not the kinds of activities engaged in by micropropagation operations, but they do demonstrate vividly the power of totipotency in plant growth.

By contrast, commercial tissue culture is the business of producing many thousands of clones of particular plants using meristematic tissue. Different plants require different formulations of media, as the formula that is ideal for one plant will produce no results with some other plants. Conventional micropropagation requires a sterile or aseptic environment and the operating costs of such a facility are substantial. However, without totipotency, non-e of this would be possible at any cost.

PLANT METABOLISM

All biological organisms have defence mechanisms to protect them from the negative effects of small quantities of foreign compounds (xenobiotics). These foreign compounds in plants include pesticides. Herbicides are a type of pesticide toxic to plants specifically, because they inhibit metabolic pathways unique to plants (*e.g.* photosynthesis). Herbicides are used successfully in weed management systems because they selectively harm sensitive weeds while leaving crops undamaged. Many factors contribute to successful herbicide action (*e.g.* successful placement or absorption and movement to target sites); however, the major reason herbicides are selective against weeds in crops is because crop plants are able to metabolize the herbicide to a non-toxic form. The basis for herbicide selectivity relies on enzymatic systems used in the plant's normal metabolic processes. The relative rates of herbicide absorption, translocation, and metabolism usually determine whether or not a herbicide will elicit a phytotoxic response. To better understand how these processes may each influence herbicidal action, they should be considered separately. Absorption has primary control over translocation, metabolism and phytotoxic action because the total amount of herbicide available for these processes is determined by the amount of herbicide absorbed by the plant. Metabolism influences both herbicide absorption and phytotoxic action. Metabolism generally converts the herbicide to a form with reduced phytotoxicity, thus, increasing the concentration gradient of the parent herbicide, so more is absorbed. Herbicide metabolism also influences phytotoxic action by either rendering the herbicide less or more active.

PHOTOPHOSPHORYLATION

Pq, the acceptor molecule, releases the excited electron into the care of an electron transport system that is sort of like a downhill bucket brigade. The transport system moves electrons extracted from water temporarily to a high-energy storage molecule called nicotinamide adenine dinucleotide phosphate (NADP). NADP is an electron acceptor for photosystem. The transport chain is essentially iron-containing pigments, cytochromes, a copper containing protein called plastocyanin and other electron transferring molecules. As electrons are passed through the chain and protons are being shuffled through a coupling factor, ATP molecules are assembled from ADP and phosphate in a process called photophosphorylation.

A similar series of events occurs in photosystem I. After a photon of light strikes a P_{700} molecule, the resulting excited electron is passed along to an iron-sulphur molecule Fe-S which in turn passes it to another acceptor molecule ferrodoxin, (Fd). The ferrodoxin molecule releases the electron to a carrier molecule called flavin adenine dinucleotide (FAD) and then eventually on to NADP. A reduction occurs and NADP+ becomes NADPH. Electrons from photosystem II and the activities of the electron transport system replace any electrons removed from the P_{700}molecule. Because the electrons move in one direction, the movement of electrons from water to photosystem II to photosystem I to NADP are said to be part of noncyclic electron flow. Any ATP that is produced are designated noncyclic phosphorylation.

It should be noted that photosystem I can operate independently of photosystem II. When this occurs, the electrons boosted from P_{700} reaction-center molecules (photosystem I) are passed through an intermediary acceptor molecule called P_{430} and then on to the electron transport chain. This is rather then to the ferrodoxin and NADP.

After being passed through the electron transport chain, the electron is dumped back into the reaction-center of photosystem I. This process demonstrates cyclic electron flow and any ATP generated by cyclic electron flow is termed cyclic phosphorylation. Note, that no water molecules are split and no NADPH or oxygen is produced.

CHEMIOSMOSIS

Earlier we mentioned in passing a coupling factor. The enzyme necessary for mediation of the splitting of water molecules is on the inside of the thylakoid membrane. As a result of this, a proton gradient forms across the membrane and the movement of these protons is thought to be a source of energy for generating ATP.

The motion is thought to be similar to molecular movement during osmosis and has hence been termed chemiosmosis. As the protons move across the membrane, they are assisted in crossing by protein channels called ATPase or coupling factor. Because of the proton movement, ADP and phosphate combine which produces ATP.

NITTY-GRITTY OF CARBON-FIXING REACTIONS

Both ATP and NADPH are important products of the light reactions and both of them play roles in the synthesis of carbohydrates from atmospheric carbon dioxide. Although the carbon-fixing reactions do not require daylight, they generally are conducted during daylight hours as there is some indication that some of the enzymes required for the processes in carbon-fixing may require some level of light. These reactions take place in the stroma of the chloroplast.

Three known mechanisms of converting carbon dioxide to sugar.

The Calvin Cycle or the 3-carbon Pathway

The Calvin cycle is the most common of the three mechanisms and has four main results:

1. With the assistance of the enzyme rubisco (RuBP carboxylase), six molecules of atmospheric carbon dioxide combine with six molecules of ribulose 1, 5-bisphosphate (RuBP)
2. The result of the first step is six unstable 6-carbon complexes, which immediately split into two 3-carbon molecules of 3-phosphoglyceric acid or 3PGA. This is the first stable compound in photosynthesis.
3. NADPH and ATP from the light-reactions, supply the energy required to convert the 3PGA to 12 molecules of glyceraldehydes 3-phosphate (GA3P), which is a 3-carbon sugar phosphate complex.
4. Finally, of the 12 molecules formed; 10 are restructured into six 5-carbon molecules of RuBP—the sugar that the process started with.

The sugars produced can either add to an increase in the sugar content (carbohydrate content) of the plant or they can be used in pathways that lead to the production of lipids and amino acids.

4-Carbon Pathway—C_4 plants

These plants use a 4-carbon molecule called oxaloacetic acid in place of the 3-carbon 3-phosphoglyceric acid used in step two of the Calvin cycle.

Oxaloacetic acid is produced from a 3-carbon compound PEP—phosphoenolpyruvate and carbon dioxide. This process is enzyme mediated and occurs in the mesophyll cells of the leaf. Some species will convert the resulting oxaloacetic acid to aspartic, malic or other acids.

Note that the acids do not substitute for 3PGA. The 4-carbon acids migrate to the bundle sheaths surrounding the vascular bundles, where they are further converted to pyruvic acid and carbon dioxide. In returning to the mesophyll cells and interacting with ATP molecules, the pyruvic acids molecules are able to produce additional PEP.

In the bundle sheath cells, the carbon dioxide formed converts into 3PGA and other molecules, by combining with RuBP. The other molecules formed are similar to the other ones formed in the Calvin cycle. The C_4 cycle furnishes carbon dioxide to the Calvin cycle in a more roundabout way than the C_3 pathway, but there is an advantage to this extra pathway.

The extra pathway greatly reduces photorespiration in C_4 plants, and this is a good thing because photorespiration is in direct competition with the Calvin cycle and even takes place in the light while the Calvin cycle is functioning. During photorespiration, RuBP reacts with oxygen to create carbon dioxide; in contrast, during photosynthesis RuBP and carbon dioxide are used to form carbohydrates.

C_4 plants are able to pick up carbon dioxide in very low concentrations via the mesophyll cells. The Calvin cycle occurs in the bundle sheath where carbon dioxide is readily available. Because of the location, the enzyme rubisco will be in a prime spot to catalyze the reaction between RuBP and carbon dioxide, rather than oxygen. As a result C_4 plants have photosynthetic rates that are two to three times higher than C_3 plants.

There are a few other characteristic features of C_4 plants worth noting:

1. C_4 plants have two types of chloroplasts and an alternate pathway for using carbon dioxide. C_3 plants only have one type of chloroplast and one pathway. Chloroplasts with starch grains and are large with very little grana, and sometimes none, in the bundle sheath cells. In the mesophyll, the small, but numerous chloroplasts have no starch grains and contain highly developed grana.
2. PEP carboxylase is found in high concentration in the mesophyll cells which permits ready conversion of carbon dioxide to carbohydrate at lower concentrations than does rubisco (in bundle sheath cells) of the Calvin cycle.
3. The temperature ranges for C_4 plants are much higher than C_3 plants which enables C_4 plants to live well in conditions that would likely kill a C_3 plant.

CAM Photosynthesis

Crassulacean acid metabolism is a modified photosynthetic system that is somewhat similar to C_4 photosynthesis in that 4-carbon compounds are produced during the carbon-fixing reactions. CAM plants accumulate malic acid in their chlorenchyma tissues at night, which is converted back to carbon dioxide during the day. In the daytime, malic acid diffuses out of the vacuoles and is converted to carbon dioxide for use in the Calvin cycle. PEP carboxylase is responsible for converting the carbon dioxide plus PEP to malic acid at night.

This modification allows for a greater amount of carbon dioxide to be converted to carbohydrate during the day than would be otherwise converted given the conditions CAM plants generally grow in. CAM plants generally close their stomata during the day in order to reduce water loss. There are more than 20 families that contain CAM plants, including cacti, stonecrops, orchids, bromeliads and many succulents growing in regions of high light intensity. There are some succulents that do not have CAM photosynthetic capabilities, as well as non-succulents that do have the ability.

There are great resources available that go into even greater detail on these reactions. If you are interested in these titles, please don't be afraid to ask on the forum for direction.

RESPIRATION

Respiration is the group of processes that utilizes the energy that is stored through the photosynthetic processes. The steps in respiration are small enzyme-mediated steps tha release tiny amounts of immediately available energy, the energy released is usually stored in ATP molecules which allow for even more efficient use of an organism's energy. Respiration occurs in the mitochondria and cytoplasm of cells.

There are several forms of respiration: aerobic—which requires oxygen, anaerobic—which occurs in the absence of oxygen, and fermentation—which also occurs in the absence of oxygen.

Aerobic respiration is the most common form of respiration and cannot be completed without oxygen gas. The controlled release of energy is the main event in aerobic respiration.

Certain types of bacteria and other organisms without oxygen gas carry on anaerobic respiration and fermentation. Compared to aerobic respiration the amount of energy released is quite small. The main difference between aerobic respiration and fermentation is in the way hydrogen is released and combined with other substances. Two very common forms of fermentation are summed up by the following equations:

$$C_6H_{12}O_6 \rightarrow \text{(with enzymes)} \rightarrow 2C_2H_5OH + 2CO_2 + \text{energy (ATP)}$$

glucose ethyl alcohol carbon dioxide

$$C_6H_{12}O_6 \rightarrow \text{(with enzymes)} \rightarrow 2C_3H_6O_3 + \text{energy (ATP)}$$

glucose lactic acid

Note the first equation is particularly valuable to the brewing industry.

MAJOR STEPS IN RESPIRATION

Glycolysis

The first step does not require oxygen gas (O_2) and takes place in the cytoplasm. The glycolytic phase is subdivided into three main steps and several smaller ones. Each step is mediated by an enzyme. A small amount of energy is released and hydrogen atoms are removed from compounds derived from glucose.

The main gist of the steps are:

A. The glucose molecules goes through several steps and becomes a double phosphorylated fructose molecule.

B. The 6-carbon fructose molecule is split into two 3-carbon fragments, each with a phosphate, GA3P

C. Hydorgen, energy and water are removed from the GA3P molecules leaving pyruvic acid.

Glycolysis requires two molecules of ATP to get the process started. In the processes, four ATP molecules are created, with a net gain of 2 ATP molecules at the end of glycolysis.

The hydrogen ions and electrons that are released are held by an acceptor molecule called NAD—nicotinamide adenine dinucleotide. The overall end products of gylcolysis is: 2-ATP molecules, 2-NADH molecules, and pyruvic acid.

The next step depends on the kind of respiration involved—aerobic, true anaerobic or fermentation.

AEROBIC RESPIRATION (WITH OXYGEN PRESENT)

The Krebs Cycle (or Citric Acid Cycle)

The Krebs cycle takes place in the fluid matrix of the cristae compartments of the mitochondria. It is called the citric acid cycle because of all the intermediate acids in the cycle. The pyruvic acid product of glycolysis is restructured, some of the CO_2 is lost and becomes acetyl CoA which then dumps into the Krebs cycle.

During the restructuring of pyruvic acid, a molecule of NADH is produced. The Krebs cycle removes energy, CO_2 and hydrogen from acetyl CoA via enzyme mediated reactions of organic acids.

The hydrogen removed during the Krebs cycle is picked up by FAD and NAD acceptor molecules. The end result of the metabolizing of two acetyl CoA molecules in the Krebs cycle is: 2-ATP molecule, oxaloacetic acid (to further drive the cycle), 6-$NADH_2$ molecules, 2-$FADH_2$ molecules and 2CO_2 molecules. The NAD and FAD molecules and the hydrogens that they carry will be dumped into the next step in respiration in order to extract the energy from the molecules.

FACTORS REGULATING RATE OF RESPIRATION

Temperature

To a point, the higher the temperature the faster respiration occurs. At some temperature, enzymes will become inactivated, although there are thermophilic (heat-loving) organisms that do quite well in high-temperature environments. Energy from sugar is released faster as the rate of respiration increases which results in a net weight loss. Plants offset the weight loss by increasing photosynthetic production of sugar. Note that during respiration, some of the energy is lost as heat, which results in an overall increase in organism temperature—not necessarily detectible by human hands.

Water

Enzymes generally operate in the presence of water, and reduced water in a plant will reduce the rate of respiration. Seeds usually have a water content of less than 10%, while mature living cells usually are in excess of 90% water. Seeds keep better if they are kept dry as the respiration rate remains quite low. However, if a seed comes into contact with water and via imbibition swells, the respiration rate will skyrocket.

The temperature could increase to the point of killing the seeds. Spontaneous combustion can occur from the respiration generated heat when a fungi or bacterium is permitted to grow on wet seeds. Kind of a neat little trivia fact to tuck away.

Oxygen—Oxygen is an important regulator of respiration. If oxygen is drastically reduced, respiration may drop off to the point of retarding growth or death. Low oxygen concentrations can lead to the onset of fermentation processes.

ASSIMILATION AND DIGESTION

Assimilation is the conversion of the sugar produced by photosynthesis to fats, proteins, complex carbohydrates and other substances. While digestion is the breakdown of large insoluble molecules by hydrolysis to smaller soluble forms that can be transported to various parts of the plant.

Summary of key differences between photosynthesis and respiration:

Photosynthesis

- Energy stored in sugar molecules
- Carbon dioxide and water used
- Increases weight
- Requires light
- Occurs in chlorophyll
- In green organisms, produces oxygen
- With light energy, produces ATP

Respiration

- Energy released from sugar molecules
- Carbon dioxide and water released
- Decreases weight
- Can occur in light or darkness
- Occurs in all living cells
- Uses oxygen (aerobic respiration)
- With energy released from sugar, produces ATP.

8

Soil–Plant pH Transpiration

Increasing acidity or alkalinity correspond to logarithmic increase in concentration of H and OH respectively (vertical bars). Horizontal bars represent relative availability (or toxicity) at any particular pH. Most agricultural soils will be slightly acid (pH around 5.5 to 6.5) and essential nutrients are all readily available within that range. Of particular note, highly acid soils are conducive to both Al and Mn toxicity and to Mo deficiency. Highly alkaline soils are conducive to B toxicity but to Fe, Zn and Mn deficiencies.

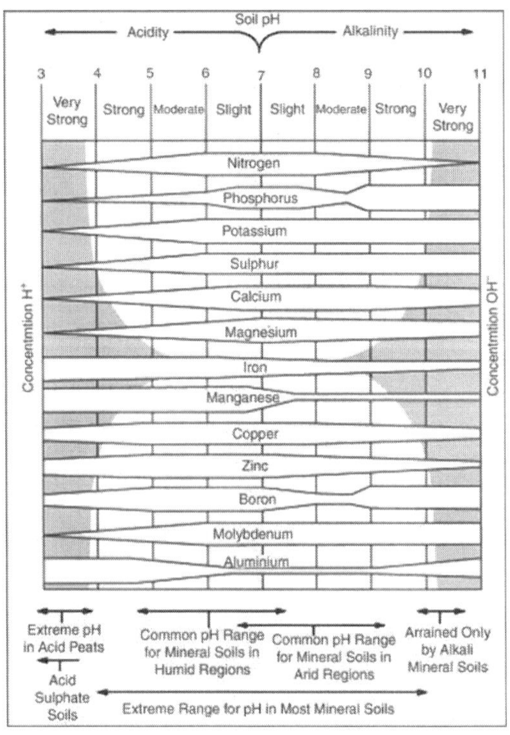

Fig. A Highly Diagrammatic Picture of Soil Nutrient Availability (and Element Toxicity) as a Function of pH.

The pH value most relevant to soil and plant chemical processes is pH of the soil solution. A soil is acidic if the pH of its aqueous solution phase is <7 and alkaline if that pH exceeds 7. Nutrient element availability varies accordingly, and beyond the range of pH 4–8 plant growth becomes a function of pH *per se*, plus pH effects on nutrient ion availability. In chemical terms, pH represents a measure of H activity in a soil solution which is in a dynamic equilibrium with a negatively charged solid phase. H ions are strongly attracted to these negative sites and have sufcient power to replace other cations from them. A diffuse layer in the vicinity of a negatively charged surface has higher H activity than the bulk soil solution.

Soil pH varies in time and space. Diurnal fluctuations of as much as one pH unit may occur, as well as spatial variations (horizontal and vertical down the soil prole). Soil pH also varies over seasons.

During seasons with low to moderate rainfall when evapotranspiration greatly exceeds precipitation, salts are not being removed by deep percolation and increased salts tend to reduce pH by forcing more of the exchangeable H ions into the soil solution. Conversely, during wet seasons, salts are removed from the topsoil and pH goes up. This season to season fluctuation in total salt content should not be confused with long-term soil acidication.

Relationships between pH and Ion Toxicity

Soil pH is a dominant influence on solubility and therefore availability and potential phytotoxicity of metals. Low pH favours free metal cations and protonated anions, higher pH favours carbonate or hydroxyl complexes. Therefore, availability of micronutrient and toxic ions (which are present in soil solution as cations) increases with increasing soil acidity. By contrast, availability of those present as anions (MoO_4, CrO_4, SeO_4, SeO_3 and $B(OH)_4$) increases with increasing alkalinity.

RHIZOSPHERE

Plant growth is dependent on availability of water and nutrients in the rhizosphere, the soil–root interface consisting of a soil layer varying in thickness between 0.1 mm and up to a few millimetres depending on the length of root hairs. Availability of nutrients in the rhizosphere is controlled by the combined effects of soil properties and interactions between plant roots and adjacent microorganisms in the surrounding soil.

Chemical conditions in the rhizosphere are usually very much different from those in the bulk soil further away from roots. Root-induced changes in the rhizosphere pH are a result of the balance between H and HCO_3 excretion, evolution of CO_2 by respiration and loss of various organic compounds known collectively as root exudates. The balance between H and HCO_3 excretion depends upon the cation/anion uptake ratio. Greater excretion of H accompanies a greater absorption of cations than anions and results in rhizosphere acidication. Conversely, when uptake of anions exceeds uptake of cations, excretion of HCO_3 exceeds that of H. The chemical form of soil N (ammonium v. nitrate) is an influential factor for the cation/anion ratio. Ammonium-fed plants take up more cations than anions, and they usually have a more acidic rhizosphere than bulk soil, while nitrate-fed plants take up more anions than cations and show the opposite relationship between rhizo-sphere and bulk soil pH. Plant effects on rhizosphere pH also vary with genotype, which can in turn influence nutrient ion availability. Overall, plants and soils must be regarded as interacting components in any ecosystem, and because plants take up more basic than acidic components, any net increase in ecosystem biomass will result in some degree of soil acidication.

TRANSPIRATION

Magnitude of Transpiration

Most of the terrestrial plants absorb water from the soil and transport it to aerial regions of the plant body. Though water is essential and an important component of living cells, the plant body loses considerable amount of water from its surfaces in the form of water vapors. The magnitude of water lost in the form of transpiration varies from plant to plant. It has been estimated that a single corn plant may transpire about 54 x 4.5 liters of water in a growing season. On the other hand, a single 16 meters tall silver maple tree can lose 54 x 4.5 liters of water per hour.

If a rough estimate can be made of an average forest in South India, 6500-8000 x 4.5 liters of water / acre / day are lost. The above figures clearly indicate the impact of transpiration, particularly in a country like India, where most of the farmer, depend on the seasonal rainfall for their agriculture. Controlling and economizing of water is one of the most important fields of research in Plant Physiology and Plant Breeding. It is also known that some plants like Potamogeton, Vallisneria (aquatic) have vestigial stomata, which are genetically controlled. If such characters are transferred to crop paints, by either plant breeding or somatic hybridization, it is possible to raise new plants which possess both high yielding and low transpiring attributes.

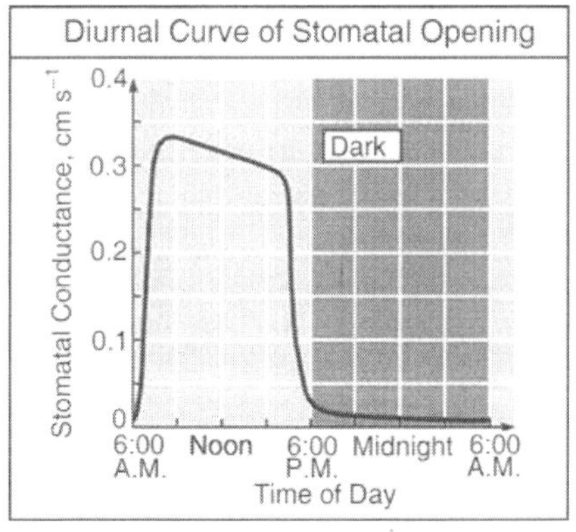

The magnitude of transpiration by plants, as mentioned before, varies among different species, and also depends upon the environmental conditions. But the most important factor that contributes to this is the structures or surfaces from which transpiration takes place.

CURTICULAR TRANSPIRATION

Cuticle is the outermost layer of varying thickness found covering the outer surface of leaves and stems etc. This layer is composed of cutin, wax and lipid which are secreted by the underlying cells. These substances are very waxy and hydrophobic in nature. However depending upon the thickness, little water may percolate or diffuse though these layers, but maximum water is lost through the breaks or openings found in the layer. Nevertheless the magnitude of transpiration is relatively insignificant.

LENTICULAR TRANSPIRATION

In this case, water is lost through lenticels, which are normally found on stems, fruits and pedicels. These structures develop during secondary growth in Dicots. These lenticels are cup like open, structures having a loose tissue called 'complementary tissue with intercellular spaces. The atmospheric air enters into the intercellular spaces of this loose tissue. If the relative humidity of the air is less, then the cells with higher water potentially lose their water into air spaces in the form of water vapors, thus water escapes into outer atmosphere. Though the magnitude of the lenticular transpiration is considerable, still it is insignificant when compared to the stomatal transpiration.

STOMATAL TRANSPIRATION

Transpiration, in this case, takes place through special pores called stomata. The amount of transpired water through these structures is more than 80-90% of the total transpired water. Hence the structure and the number of these stomata play a significant role in the water.

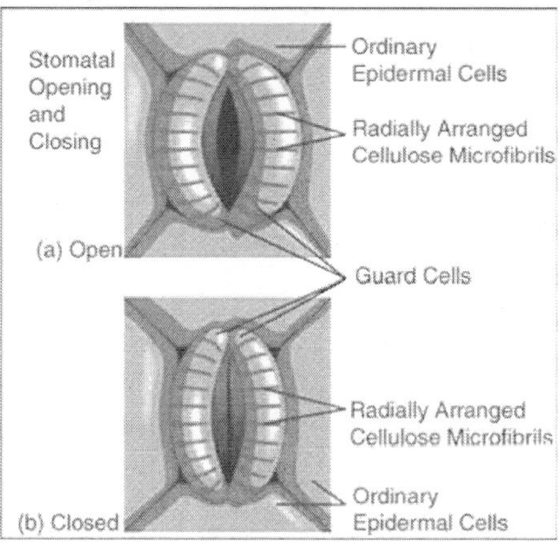

Fig. Stomatal Transpiration

Most of the stomata are found in leaves, but the presence of these structures on stems, petioles, sepals and petals is not uncommon; the numerical distribution of the stomata in a given leaf can be calculated by stomatal index. Stomatal index is a measure of total number of stomata per total number of epidermal cells in a given unit area. Different plants show different stomatal index. Further the number varies not only from plant to plant, but also varies in different environmental conditions.

Basing on stomatal index different stomatal types have been recognized:

- Apple type Mulberry type
- Potato type,
- Ota type,
- Water lily type
- Potomogeton type.

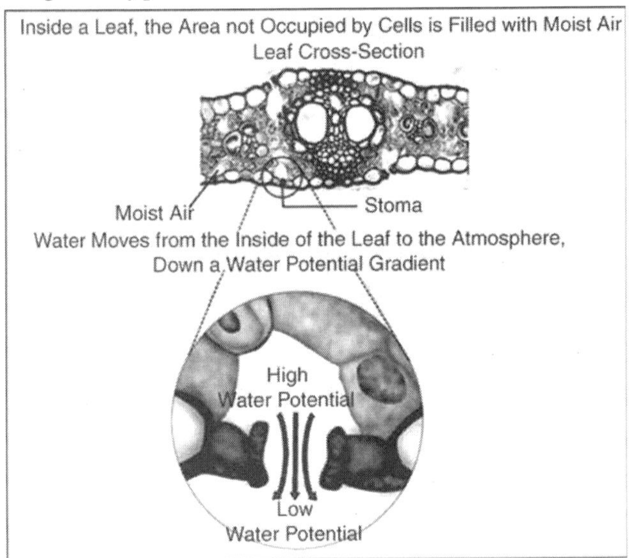

Fig. Water Moves from the Inside of the Leaf to the Atmosphere

Efficiency of Stomata

As stomata are small pores, water evaporates efficiently. When the perimeters of all the pores of stomata are taken together and then compared with the total surface area, the former far exceeds the latter. Added to this the water escapes in higher magnitude at perimeters, than the central region of the pores. Hence stomatal apparatus is well suited for higher transpiration rate.

Stomatal Apparatus

The stomatal apparatus of monocots and dicots vary in their structure and dimensions. Generally, the stomatal apparatus is derived from epidermal cells. It consists of a pore, and it is surrounded by two kidney shaped epidermal cells called guard cells, which in turn are surrounded by subsidiary cells or accessory cells. The guard cells are also in contact with spongy parenchymatous cells of inner mesophyll. Normally stomata lead into relatively large air spaces. Guard cells in the case of monocots are dumbbell shaped structures with thick walls on both face and thin wall and thin wall at the inflated ends.

They are also surrounded by subsidiary cells. The guard cells are unique in having the following features; they have active nucleus; their inner walls are considerably thick and the outer walls are very thin and extensible the ends of the paired guard cells are tightly concerned.

There are distinct cytoplasmic strands or plasmodesmata connections between inner spongy parenchymatous cells and guard cells. The most important feature of these cells is the presence of very well developed and photosynthetically active chloroplasts which are totally absent in other epidermal cells. These guard cells are also highly sensitive to the changes in pH, CO_2, light, temperature, phytohormones, and water stress.

GENERAL MECHANISM OF STOMATAL TRANSPIRATION

Water absorbed by the root system is transported upwards through xylem elements by transpiration pull mechanism and reaches the serial parts like leaves. Through the vascular system found in the veins of leaves, the water enters into mesophyll cells. If the relative humidity of the atmosphere present in air spaces is low, water from the surrounding cell surfaces, escapes in the form of water vapors into these spaces.

If the stomata are opened, water vapors escape into air, or remain in the spaces if the stomata are closed. Thus stomatal movement plays a pivotal role in regulating the loss of water from the plants.

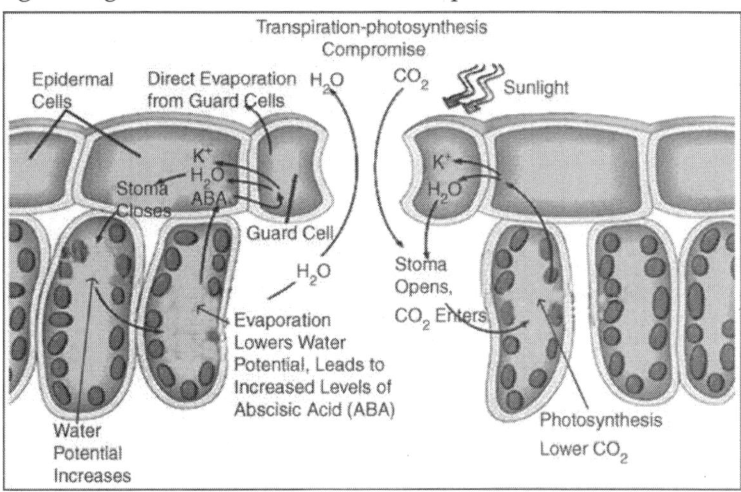

Fig. Transpiration-Photosynthesis Compromise.

The Mechanism of Opening and Closing of the Stomata

Though this process is known to Plant Biologists for more than a century, there is no unequivocal explanation which explains the mechanism of opening and closing of stomata.

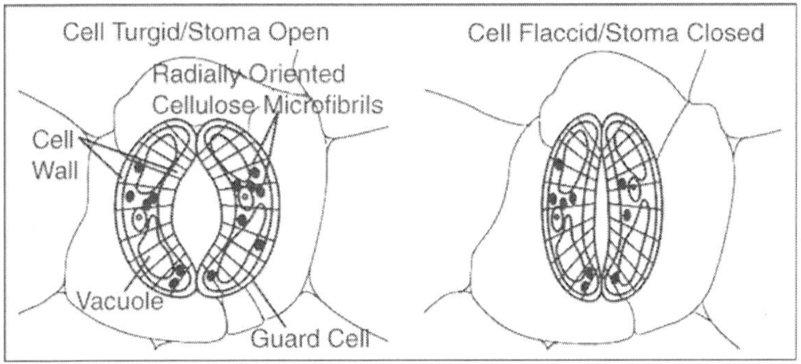

Fig. Change in Guard Cell Shape and Stomatal Opening and Closing

This is because of the intrinsic and extrinsic factors which play a major role in the guard cell movements. Nevertheless plant physiologists have been proposing theories from time to time, but they are either being contradicted or lagging in convincing explanation with reference to some specific experimental observations. Despite its complexity, it is worthy to make an attempt to understand the mechanism of opening and closing of stomata.

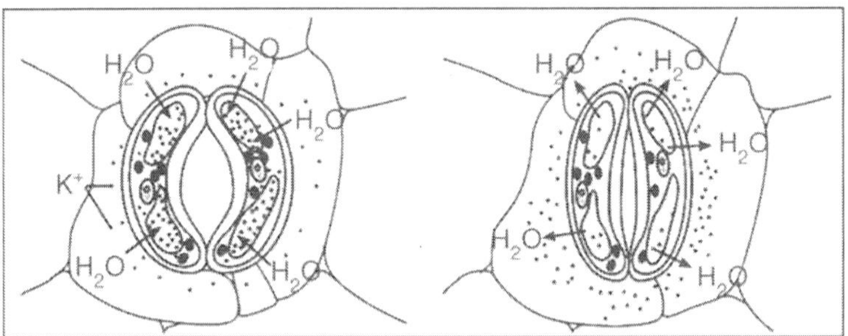

Fig. Role of Potassium in Stomatal Opening and Closing

It is interesting to know, how stomata respond to various factors and what effects these factors have on guard cells:

- Stomata open at day time and remain closed during night times. This means that light plays an effective role.
- Blue and red light of solar electro-magnetic radiation spectrum favours opening which is incidentally the action spectrum of photosynthesis. Other wave lengths of light have no significant effects.
- Increased CO_2 concentration induces the closing, and decreased concentration stimulates the opening of the stomata.

- Stomata begin to open when the pH of the medium is high (alkaline) and close when pH is low (acidic).
  - During day time Starch disappears in guard cells and at night times Starch accumulates. This feature is contrary to the other mesophyll cells, where Starch disappears at nights and accumulates during day times.
- Influx of K ions into guard cells induces the opening and efflux of K favours closing
- Phytohormone like Cytokinin stimulates the guard cells to open and Abscissic Acid (ABA) induces the closing.
- In succulents like cactus etc., stomata open during nights and close at day times. This is correlated to Carboxylic Acid Metabolism (CAM) where organic acids accumulate during night and the same are decarboxylated at day time. The notable organic acids synthesized are mainly Malate and Glycollate.
- Glycollate induces the opening of stomata and glycollate oxidation favours the closing of the stomata.

THE ELEMENTS OF COMPLETE PLANT NUTRITION

Complete nutrition results in superior plant growth. Why choose anything less for your plants? There are 20 elements necessary for optimum plant growth. Air and water supply carbon, hydrogen and oxygen. Macronutrients are required by plants in large amounts. Micronutrients are required in smaller amounts. Eliminate any of these elements,and plants will display abnormal growth, deficiencies or may not reproduce. The following is a brief guide to the role played by each of these essential nutrient elements.

Macronutrients

- *(N)-Nitrogen*: Component of proteins, hormones, chlorophyll, vitamins and enzymes. Promotes stem and leaf growth. The ammoniacal and nitrate forms are used directly by plants for stem and leaf growth. The urea form of nitrogen must be broken down by soil borne microorganisms or urease before it can be utilized by the plant. Urea can cause leaf tip and root burn. Deficiency (Def.): reduced yields, yellowing of leaves, stunted growth. Excess nitrogen can delay fruiting and flowering.
- *(P)-Phosphorus*: Seed germination, photosynthesis, protein formation, overall growth and metabolism, flower and fruit formation. Def.: purple stems and leaves, retarded growth and maturity, poor flowering and fruiting. Large amounts without zinc cause zinc deficiency. Low pH (<4) ties up phosphates in organic soils.
- *(K)-Potassium*: Formation of sugars, carbohydrates, proteins, cell division. Adjusts water balance; improves stem rigidity and cold hardiness; enhances flavour, colour and oil content of fruits;

important for leafy crops. Def.: spotted, curled or burned look to leaves; lower yields.

- (*Ca*)-*Calcium*: Activates enzymes; structural part of cell walls; influences water movement, cell growth and division Required for uptake of nitrogen and other minerals. Leached from soil by watering. Immobile-requires a constant supply for growth. Def.: stunting of new growth in stems, flowers, roots; black spots on leaves and fruit; yellow leaf margins.
- (*Mg*)-*Magnesium*: Critical component of chlorophyll; needed for functioning of enzymes for carbohydrates, sugars and fats; fruit and nut formation; germination of seeds.
- Def.: yellowing between veins of older leaves; chlorosis; leaf droop. Leached by watering. Foliar spray to correct deficiencies.
- (*S*)-*Sulfur*: Component of amino acids, proteins, vitamins, enzymes. Essential for chlorophyll. Imparts flavour to many vegetables. Def.: light green leaves. Water supply may contain sulfur. Leached by watering.

Micronutrients

- (*B*)-*Boron*: Affects at least 16 functions: flowering, pollen germination, fruiting, cell division, water relationships, movement of hormones, cell wall formation, membrane integrity, calcium uptake, movement of sugars. Immobile; easily leached. Def.: terminal bud die back causes rosette of thick, curled, brittle leaves or brown, discolored, cracked fruits, tubers and roots.
- (*Cl*)-*Chlorine*: Involved in osmosis (movement of water or solutes in cells), ionic balance necessary to take up mineral elements and photosynthesis. Def.: wilting, stubby roots, yellowing, bronzing. Scents in some plants may be decreased. Leached by watering.
- (*Co*)-*Cobalt*: Required by nitrogen fixing bacteria; formation of B12 vitamin; formation of DNA. Will extend life of cut flowers such as roses. Def.: may result in nitrogen deficiency.
- (*Cu*)-*Copper*: Necessary for nitrogen metabolism; component of enzymes - may be part of enzyme systems that use carbohydrates and proteins. Bound tightly in organic matter. May be deficient in highly organic soils. Not readily lost from soil but may be unavailable. Def.: die back of shoot tips; terminal leaves develop brown spots. Excess is toxic.
- (*Fe*)-*Iron*: Enzyme functions; catalyst for synthesis of chlorophyll; essential for new growth. Def.: pale leaves, yellowing of leaves and veins. Leached by water and held in lower parts of soil. High pH soils may have iron present but unavailable to plants.
- (*Mn*)-*Manganese*: Enzyme activity for photosynthesis, respiration and nitrogen metabolism. Def.: young leaves are pale with green veins

similar to iron deficiency; advanced stages-leaves are white and drop; brown, black or gray spots may appear next to veins. Plants in neutral or alkaline soils often show def. Acid soils may increase uptake causing toxicity.

- (*Mo*)-*Molybdenum*: Structural part of enzymes that reduce nitrates to ammonia for amino acid development essential to protein formation; required by nitrogen fixing bacteria. Def.: pale leaves with rolled, cupped margins. Seeds may not form. Nitrogen deficiency may occur if plants are lacking Mo.
- (*Ni*)-*Nickel*: Recently recognized as essential. Required for the urease enzyme to break down urea into usable nitrogen and for iron uptake. Seeds require nickel to germinate.
- (*Na*)-*Sodium*: Improves nitrogen metabolism in many plants, involved in osmotic (water movement) and ionic balance in plants. Def.: yellowing of leaves and leaf tip burn; may inhibit flower formation.
- (*Si*)-*Silicon*: Component of cell walls; enhances resistance to sucking insects and fungi.Foliar sprays reduce populations of aphids on some plants. Enhances leaf presentation; improves heat, drought and cold tolerance; improves photosynthesis; extends bloom life. Def.: wilting, poor fruit and flower set, increased susceptibility to insects and disease. Disease resistance is enhanced by regular foliar feeding.
- (*Zn*)-*Zinc*: Functional part of enzymes including auxin (growth hormone) synthesis, carbohydrate metabolism, protein synthesis, stem growth. Def.: mottled leaves, irregular yellow areas. Zinc deficiency leads to iron deficiency. Occurs in eroded soils; least available at pH of 5.5-7.0. Lower pH can cause availability to the point of toxicity.

NUTRIENT BALANCE AND NUTRIENT DEFICIENCIES

Plants (like people) need a 'balanced diet'. They need all 13 nutrients to remain healthy. If one is missing, the plant will not grow well. Poor plant nutrition causes plants to grow slowly in the nursery and in the field, and to be more susceptible to diseases. Many people confuse the symptoms of nutrient deficiencies with those of too much or not enough shade or water.

In fact, all three factors, shade, water and nutrients affect plant growth, and interact to produce healthy plants. A plant that grows in full light with abundant moisture and receives all the 13 nutrients will grow fast and have a dark green colour in its leaves. A plant that grows slowly in the shade may also have dark green leaves, but when exposed gradually to the sun, the leaves may turn yellow.

This does not mean that plants do not like full sun—it might indicate a nutrient deficiency which did not show up in the shade because the plant did not have enough light to stimulate fast growth. Together, water, shade and nutrients must be monitored and adjusted to produce quality seedlings. It takes practice to learn the signs identifying a missing nutrient or nutrients, but you can learn to do so, and some of the signs are common to many plants. A *good nursery practice* is to carefully monitor the leaves of your plants for signs of nutrient deficiency, and correct them with a better substrate or with fertilizer.

Inorganic Fertilizers

Inorganic fertilizers are mined from the soil, or produced during complicated chemical reactions. A *good nursery practice* is to read the fertilizer labels. This allows you to apply what the plants need without wasting nursery resources. Fertilizers contain only plant nutrients; they are not used to combat plant diseases or insects. Inorganic fertilizers do not improve the substrate physical properties, whereas organic material such as compost does. Inorganic fertilizers are also expensive and not always available in the stores.

Nursery managers should carefully consider the cost and benefit of buying these products. Granular fertilizers are commonly given names like "17-17-17", or "10-30-10". What do the numbers mean? They represent the percentages of nitrogen (N), phosphorus (P), and potassium (K) in the fertilizer—17% N, 17% P, 17% K. In this case, 51% of the mixture is made up of N-P-K, and the rest is inactive material used to help spread the fertilizer evenly. Urea contains only N, and is labelled as 46-0-0. Urea is very strong and can easily burn the plants if too much is applied.

Granular Fertilizers

Granular fertilizers can be mixed into the substrate or into the irrigation water, or be applied to older plants on the soil surface. It is better to mix the fertilizer directly into the substrate before planting the seed because the roots can avoid or seek the fertilizer as they need it. Use only small quantities such as 2 or 4 grams (1/2 teaspoon) per 1 kg of soil. It is better to add too little then too much. You need to experiment with different levels. Plants should respond within two weeks. When dissolving fertilizer in warm water, carefully note whether it is thoroughly dissolved. If not, it is probably the phosphorus that remains. It may be better to apply fertilizer in granular form if it does not dissolve thoroughly. Apply liquid fertilizer to the soil, not to the leaves which are easily burnt if fertilizer remains on them. Be extremely cautious when applying fertilizer to young plants.

Foliar Fertilizers

Foliar fertilizers are used in order to get the nutrients to the plants quickly. They are specially formulated to apply directly on the leaves. Foliar fertilizers are absorbed by the leaves, not by the roots. When plants are acutely deficient in nutrients, foliar fertilizers often help 'green them up'. Frequently, foliar fertilizers only contain the micronutrients, since it is assumed that the macronutrients are available in the substrate.

However, some such as 'GroGreen'® contain both micronutrients and 20-30-10 of N-P-K. Often an adhering agent such as 'Da-Plus' is used to help the fertilizer stay on the leaves so that it is not washed off in the rain. Because foliar fertilizers are expensive, and they may not encourage strong root growth, they should not be used as a long-term solution for plant nutrients.

COMMON NUTRIENT DEFICIENCY SYMPTOMS OF MACRONUTRIENTS

- *Nitrogen*: This is a mobile nutrient, which means that when nitrogen is deficient, plants move it from the older foliage to the younger, actively growing leaves. The older leaves (the ones lower on the stem of the tree) become yellow first, while the new leaves remain green.
- *Phosphorus*: The entire seedling is stunted, especially during early growth. Depending on the species, the leaves may become dull green, yellow or purpletinged. The purpling of leaves is a classic symptom, but sometimes there are no colour differences in leaves, so visual diagnosis is not always reliable. The purple colour should not be confused with new leaves that often appear purple or red when they first flush out.
- *Potassium*: Symptoms appear in older leaves first. These start to yellow at the edges, and have some green at the base. Later, leaf edges turn brown and may crinkle or curl and small necrotic (dead) spots may appear. Plants may wilt, even though sufficient water is available in the substrate. When deficiencies are severe, leaves will die. Calcium: This is difficult to detect because signs include slow growth, and dieback of bud or root tips. Seedlings will have stubby little roots with brownish discoloration. The problem is most common in very acidic soils. A well-developed root system with many fine root hairs is important for calcium uptake.
- *Magnesium*: This nutrient is commonly deficient in coarse-structured soils and in acidic soils. Uptake may be blocked if there is too much potassium in the soil. Like nitrogen, magnesium is a mobile nutrient, so deficiency symptoms show up in the older leaves first. These leaves show a very characteristic yellowing between the veins or ribs, and they appear streaked.
- *Sulphur*: Plants will be slightly stunted. This is not a mobile nutrient, so the symptoms show up on younger leaves which are initially light green, but eventually develop scorched and curled margins. Dry areas can form along the margins and then spread inward to the leaf midrib.

NATURAL SOURCES OF SOIL NITROGEN

The nitrogen in soil that might eventually be used by plants has two sources–nitrogen containing minerals and the vast storehouse of nitrogen in the atmosphere. The nitrogen in soil minerals is released as the mineral decomposes. This process is generally quite slow and contributes only slightly to nitrogen nutrition on most soils.

On soils containing large quantities of NH_4 rich clays (either naturally occurring or developed by fixation of NH_4 added as fertilizer), however, nitrogen supplied by the mineral fraction may be significant in some years. Atmospheric nitrogen is thought to be a major source of nitrogen in soils. In the atmosphere it exists in the very inert N_2 form and must be converted before it becomes useful in the soil. This conversion is accomplished two ways. Some N_2 is oxidized to NO_3 by lightning during thunderstorms.

The NO_3 dissolves in raindrops and falls into the soil. The quantity of nitrogen added to the soil in this manner is directly related to thunderstorm activity, but most areas probably receive no more than 20 lb nitrogen/acre per year from this source.

Some microorganisms can utilize atmospheric N_2 to manufacture nitrogenous compounds for use in their own cells. This process, called biological nitrogen fixation, requires a great deal of energy; therefore, free-living organisms that perform the reaction, such as *Azotobacter*, generally fix little nitrogen each year (usually less than 20 lb nitrogen/acre), because food energy is usually scarce. Most of this fixed nitrogen is released for use by other organisms upon death of the microorganism.

Bacteria such as *Rhizobia*, that infect (nodulate) the roots of, and receive much food energy from, legume plants can fix much more nitrogen per year (some well over 100 lb nitrogen/acre). When the quantity of nitrogen fixed by *Rhizobia* exceeds that needed by the microbes themselves, it is released for use by the host legume plant. This is why well-nodulated legumes do not often respond to additions of nitrogen fertilizer. They are already receiving enough from the bacteria.

Nitrogen Transformations and Losses in Soils

Nitrogen can go through many transformations in the soil. These transformations are often grouped into a system called the "nitrogen cycle", which can be presented in varying degrees of complexity. Because microorganisms are responsible for most of these processes, they occur very slowly, if at all, when soil temperatures are below 50 °F, but their rates increase rapidly as soils become warmer. The heart of the nitrogen cycle is the conversion of inorganic to organic nitrogen, and vice-versa. As microorganisms grow, they remove NH_4 and NO_3 from the soil's inorganic, available-nitrogen pool, converting it to organic nitrogen in a process called immobilization.

When these organisms die and are decomposed by others, excess NH can be released back to the inorganic pool in a process called mineralization. Nitrogen can also be mineralized when microorganisms decompose a material containing more nitrogen than they can use at one time, materials such as legume residues or manures. Immobilization and mineralization are conducted by most microorganisms, and are most rapid when soils are warm and moist, but not saturated with water.

The quantity of inorganic nitrogen available for crop use often depends on the amount of mineralization occurring, and the balance between mineralization and immobilization. Ammonium ions (NH_4) not immobilized or taken up quickly by higher plants are usually converted rapidly to NO_3 ions by a process called nitrification. This is a two step process, during which bacteria called Nitrosomonas convert NH_4 to nitrite (NO_2), and then other bacteria, Nitrobacter, convert the NO_2 to NO_3. This process requires a well aerated soil, and occurs rapidly enough that one usually finds mostly NO_3 rather than NH_4 in soils during the growing season. The nitrogen cycle contains several routes by which plant-available nitrogen can be lost from the soil. Nitrate-nitrogen is usually more subject to loss than is ammonium nitrogen. Significant loss mechanisms include leaching, denitrification, volatilization, and crop removal. The nitrate form of nitrogen is so soluble that it leaches easily when excess water percolates through the soil. This can be a major loss mechanism in coarse-textured soils where water percolates freely, but is less of a problem in finer-textured, more impermeable soils, where percolation is very slow.

These latter soils tend to become saturated easily, and when microorganisms exhaust the free oxygen supply in the wet soil, some obtain it by decomposing NO_3. In this process, called denitrification, NO_3 isconverted to gaseous oxides of nitrogen or to N_2 gas, both unavailable to plants. Denitrification can cause major losses of nitrogen when soils are warm and remain saturated for more than a few days. Losses of NH_4 nitrogen are less common and occur mainly by volatilization. Ammonium ions are basically anhydrous ammonia (NH_3) molecules with an extra hydrogen (H) attached. When this extra H is removed from the NH_4 ion by another ion such as hydroxyl (OH), the resulting NH_3 molecule can evaporate, or volatilize from the soil. This mechanism is most important in high pH soils that contain large quantities of OH ions. Crop removal represents a loss because nitrogen in the harvested portions of the crop plant are removed from the field completely. The nitrogen in crop residues is recycled back into the system and is better thought of as immobilized rather than removed. Much is eventually mineralized and may be reutilized by a crop.

Nitrogen Needs of and Uptake by Plants

Plants absorb nitrogen from the soil as both NH_4 and NO_3 ions, but because nitrification is so pervasive in agricultural soils, most of the nitrogen is taken up as nitrate. Nitrate moves freely towards plant roots as they absorb water. Once inside the plant NO_3 is reduced to an NH_2 form and is assimilated to produce more complex compounds. Because plants require very large quantities of nitrogen, an extensive root system is essential to allowing unrestricted uptake. Plants with roots restricted by compaction may show signs of nitrogen deficiency even when adequate nitrogen is present in the soil.

Table. Utilization of Nutrients by Various Crops.

Crop	Yield Per	N	P_2O_5	K_2O	Mg	S	
Alfalfa	8 tons	450	80	480	40	40	
Corn	180 bu.	240	100	240	50	30	
Coastal	10 tons	500	140	420	50	40	Bermuda
Soybeans	60 bu.	324	64	142	27	25	
Wheat	80 bu.	134	54	162	24	20	

Most plants take nitrogen from the soil continuously throughout their lives and nitrogen demand usually increases as plant size increases. A plant supplied with adequate nitrogen grows rapidly and produces large amounts of succulent, green foliage. Providing adequate nitrogen allows an annual crop, such as corn, to grow to full maturity, rather than delaying it.

A nitrogen-deficient plant is generally small and develops slowly because it lacks the nitrogen necessary to manufacture adequate structural and genetic materials. It is usually pale green or yellowish, because it lacks adequate chlorophyll. Older leaves often become necrotic and die as the plant moves nitrogen from less important older tissues to more important younger ones. On the other hand, some plants may grow so rapidly when supplied with excessive nitrogen that they develop protoplasm faster than they can build sufficient supporting material in cell walls. Such plants are often rather weak and may be prone to mechanical injury. Development of weak straw and lodging of small grains is an example of such an effect.

ORGANIC NITROGEN IN SOILS

The supply of available nitrogen in soils is often supplemented by nitrogen released from soil organic matter or organic materials added to soils (manure, residues of forage legumes, etc.). The quantity of nitrogen released from any of these materials depends on the composition of the material, particularly its ratio of carbon to nitrogen, and the weather. When microorganisms decompose materials such as manure and alfalfa residue, that have C/N ratios of 10:1 to 20:1 (lb carbon per lb nitrogen), net mineralization occurs, because the microbes are releasing nitrogen from the residue faster than they can utilize it.

When they are decomposing residues such as cornstalks or wheat straw, that have much larger C/N ratios, they must take some extra mineral nitrogen from the soil to utilize all of the carbon they are digesting, so net immobilization occurs. The rate at which these processes occur depends on weather and soil conditions. When soils are warm and moist, decomposition proceeds rapidly, and nitrogen released from legume residues or manures may be significant, but when soils are cold and wet, or very dry, nitrogen release may be very much less than expected.

There are situations in which nitrogen applications can be reduced or eliminated when corn follows good stands of alfalfa or another legume. However, the nitrogen relationships involved here are so site-specific and weather-related that local recommendations should be consulted before making decisions regarding nitrogen credits. Legume species, stand density, age, soil drainage, and timing of operations are all factors to be considered.

The timing and rate of manure application should also be considered when calculating nitrogen credits, and again, due to influences of weather and soil, local guidelines should be used.

Manure can be applied at rates that supply all nitrogen, phosphorus, and potassium to a crop, but such rates usually cause significant over-applications of phosphorus and potassium. Over-applications of phosphorus are important water quality concerns in many regions, so manure applications should be at rates low enough to avoid this over-application. At such low application rates, supplementary nitrogen-fertilizer applications are usually needed in crops such as corn.

COMMERCIAL NITROGEN FERTILIZERS

At one time the fertilizer industry relied on organic or direct-mined materials as sources of nitrogen. Today, virtually all nitrogen materials are manufactured, usually from ammonia. Such materials are less expensive, more concentrated, and just as plant available as the organics used in the past.

Readily Soluble Materials

Anhydrous Ammonia (NH_3) and Aqua Ammonia

Anhydrous ammonia is 82% nitrogen and is a gas at normal atmospheric temperatures and pressure. It is stored as a liquid under pressure and is injected several inches below the soil surface, where it vaporizes and dissolves in the soil water to form NH_4 ions. Ammonia (not ammonium) is harmful or toxic to living organisms, and applications near living plants may cause temporary injury. Applications too close to seeds or seedlings may cause stand problems. Though some soil sterilization occurs in the immediate zone of ammonia application, the effect is temporary, and soil life is restored in the zone within a few weeks of application. Aqua ammonia, a solution of ammonia in water (usually 21% nitrogen), is sometimes used as a fertilizer. It is also injected underground, due to the ammonia's tendency to volatilize from the solution.

Ammonium Nitrate (NH_4NO_3)

Ammonium nitrate is usually used as a solid material with an analysis of up to 34% nitrogen. It contains both NH_4 and NO_3-forms of nitrogen, and is used as a source of nitrogen in many blends of liquid and dry fertilizers, as well as being applied directly. Pure ammonium nitrate is very hygroscopic and can be explosive under certain conditions; however, present fertilizer grades of the material are specially conditioned, and when stored and handled properly, pose no problem or hazard.

Urea $[CO(NH_2)_2]_2$

Urea is a manufactured, organic compound containing 46% nitrogen that is widely used in solid and liquid fertilizers. It has relatively desirable handling and storage characteristics, making it the most important solid nitrogen-fertilizer material, worldwide. It may contain small concentrations of a toxic decomposition product, biuret; however, urea manufactured using good quality control practices rarely contains enough to be of agronomic significance. Urea is converted to ammonium carbonate by an enzyme called urease when applied to soil. Ammonium carbonate is an unstable molecule that can break down into ammonia and carbon dioxide. If the ammonia is not trapped by soil water, it can escape to the atmosphere. This ammonia volatilization can cause significant losses of nitrogen from urea when the fertilizer is applied to the surface of warm, moist soils, particularly those covered with plant residues (no-till) or those drying rapidly. Relatively high surface pH also aggravates nitrogen volatilization from urea.

Non-pressure Nitrogen Solutions

These materials are generally composed of urea and/or ammonium nitrate dissolved in water. The most popular are urea-ammonium-nitrate (UAN) solutions. Nitrogen concentrations generally range from 19-32%. These are very versatile materials and are often used as carriers for herbicides or applied in irrigation water. They can be injected, or banded on the soil surface, to minimize potential volatilization losses of the urea component in no-till or surface application situations. Spraying over live foliage can cause severe tissue burn, but the problem can often be minimized by applying in narrow streams rather than by broadcasting. Because the solubility of the nitrogen salts in these solutions decreases as temperature declines, these materials are subject to "salting out". This limits storage and use of more concentrated formulations in cooler seasons and regions.

Ammonium Sulfate $[(NH_4)_2SO_4]$

Ammonium sulfate is a by-product of many industrial processes and contains 20% nitrogen and 24% sulfur. It is not a widely used material, due to its relatively low nitrogen content, but may become more popular as sulfur deficiencies become more wide spread. It is not as prone to nitrogen volatilization as urea, but is slightly more sothan ammonium nitrate, particularly on alkaline soils.

Ammonium Phosphates

Ammonium phosphates are manufactured in a wide variety of forms and formulations including mono-ammonium phosphate (MAP), diammonium phosphate (DAP) and several ammonium polyphosphates. These materials are used mainly as carriers for phosphorus, but can also be significant in nitrogen management programmes, particularly when they are used as starter and row-applied fertilizers. When used in a starter programme, DAP (at any rate) is usually placed several inches from the seed to avoid damage due to any ammonia released as the material dissolves.

Nitrate Salts

Several salts of nitrate, including sodium nitrate ($NaNO_3$, 16% nitrogen), calcium nitrate ($Ca(NO_3)_2$, 15.5% nitrogen), and potassium nitrate (KNO_3, 14% nitrogen) are available, but are not widely used due to relatively high cost and low analysis. They are used mainly in specialty situations, such as production of horticultural crops.

Slow-Release Materials

There are a number of materials available that release nitrogen slowly into the soil rather than very shortly after application as do those listed above. They are useful in situations where the producer does not want high levels of soluble nitrogen in the system at any one time, due to potentials for excessive losses or excessively rapid plant growth.

Such situations occur in container-based horticultural production and inturfgrass management, among others. Inorganic compounds, such as magnesium ammonium phosphate, and organic compounds, such as urea-formaldehyde (urea form) and isobutylidene diurea (IBDU),are often used to supply nitrogen at slow but fairly predictable rates. Nitrogen release from these compounds is slowed by very low solubility in water, and release from the organic materials is usually related to rate of microbial degradation. Simpler nitrogen compounds, such as urea, can be coated with less soluble materials, including polymers and waxes, and nitrogen release controlled by the rate at which the coating deteriorates. Sulfur-coated urea (SCU), made by spraying urea granules with molten sulfur, is a common example of such a product, though more are becoming available. All of these products are relatively expensive sources of nitrogen, and their use is restricted to specialty markets and use on high-value crops.

RATE OF DIRECTION OF TRANSPORTATION

The rate with which the water is transported along the length of the stem varies from plant of plant. External conditions also play a significant role in controlling the rate of ascent of sap. But, under normal conditions the rate is 75-100 cm/hr. This is quite a rapid process. Generally most of the water is translocated upwards *i.e.* in longitudinal direction, but some of the water is also translocated horizontally to reach the peripheral tissues.

VITAL THEORIES

These theories are mostly based on the assumption that living cells play a vital role in pushing or pumping the water upwards. Westermaier (1883-1884), Godlewski (1884), Janse (1887) and others thought that the xylem vessels through which water is transported would just act as vessels through which water is transported would just act as reservoir. In this method both living and nonliving xylem forms an integrated system and act co-ordinatingly in the transport of water upwards. Unfortunately these theories had little experimental evidence to substantiate their claims. Sir J.C. Bose (1928-29) the elder brother of Subash Chandra Bose proposed an interesting theory called 'Pulsation theory.

His theory was based on the assumption that living cells all-round the xylem tissues are in a kind of rhythmic contraction and expansion, similar to that of heart in animals. This rhythmic pulsation is responsible for the movement of an instrument 'Cresograph where an electrode is inserted deep into the plant stem and the pulsations were graphically recorded.

The oscillations found on the graph indicated the pulsation activity of the cells. A little later Molish (1928-29) provided the plant with some heart stimulant and demonstrated an increased pulsation activity. On the contrary anesthetics brought down the pulsation rhythm.

Recent investigations however, using respiratory inhibitors like DNP, KCN, showed that the synthesis of ATP is essential for the rapid movement of water in the xylem parenchyma and other islets of living cells intermixed with the longitudinal array of trachieds and tracheae. These strongly support the view that vital activity, in the sense, the energy is required for the ascent of sap.

Root Pressure Theory

Absorption of water by roots has been mainly a passive process. However, the involvement of an active process is not ruled out totally. On a rainy day, when the atmospheric humidity is at its maximum, and transpiration is at its minimum, root system absorbs excess of water than it can normally absorb. As a result of it hydrostatic pressure is built up within the roots, and this is called root pressure. This is believed to act as the motive force to force the water into the xylem columns upwards.

Under the above said environmental conditions, water is forced out of the water-stomata as guttated water. Hence root pressure has been considered as an important phenomenon in ascent of sap. However, it has been noted that, some of the tallest trees found on this planet do not show any root pressure. Thus this theory fails to explain the transportation of water especially in tall trees.

Passive or Physical Force Theories

Physical forces like capillary force, collision force, atmospheric pressure, imbibitions, diffusion pressure, are found to operate in plants in one way or the other. Along with the development of science of plant physiology, people from time to time have come out with various theories involving one or to time have come out with various theories involving one or the other physical force as an explanation for ascent of sap.

Atmospheric Pressure Theory

The protagonists of this theory have assumed that plants are closed systems. When water escapes by transpiration from the surface of the leaves, it is believed that vacuum will be created within the plant body. As the root system is submerged in soil water, with the atmospheric action on the soil water, in order to fill up the vacuum created in the xylem vessels, water just enters passively; thus the water is translocated upwards.

Unfortunately plants are not closed systems but they exhibit openness, for, the gases can diffuse into and out of the plant system with ease and facility. Added to this, atmospheric pressure can support and facility. Added to this, atmospheric pressure can support the water to be lifted only to a height of 34 feet; but there are plants which are taller than this and still there is transport of water. Hence it can be concluded that atmospheric pressure could not be the force for ascent of sap.

CAPILLARY FORCE THEORY

When one end of the blotting paper or a chalk piece is dipped into ink, the ink slowly moves up. This movement through the paper is called capillary movement. Blotting paper is made up of innumerable cellulose fibres interwoven into a close network. Between such fibres, extremely narrow spaces are found, which are connected with each other and form a fine net work of capillary canals. If water is provided to such capillary system at one end, water is sucked in and it moves along the channels of capillary network by a force called capillary force.

According to capillary network by a force theory, such capillary system exists within the plant body. Tracheids and tracheae which are found longitudinally oriented in the vasculature have lumen as empty space, roots to terminal regions of the stem as continuous capillary system. When water is absorbed by the root system, the capillary system of xylem elements take up the water by capillary force and the water is supported to move upwards slowly but steadily.

Though the network xylem elements can be compared to capillary system, the rate of movement of water in a capillary system is extremely slow in comparison to the actual rate of ascent of sap observed. Furthermore the lumen of larger number of tracheids and tracheae has diameters greater than the large capillary spaces. Though capillary system may be contributing some force for the movement of water upwards, it cannot be considered as the sole force for the movement of water.

IMBIBITION THEORY

Dry wood when soaked in water swells; wooden shutters in rainy season are difficult to close; seeds soaked in water overnight get swollen. In all the above said instances the volume of the wood or seeds increases. This is due to the intake of water into the dry plant material. Wood is made up of cellulose, hemicelluloses and lignin. All these substances are hydrophilic or water loving in nature. When such materials come in contact with water, the hydrophilic substances imbibe water, which gets absorbed onto them.

Because of the adsorption of water, the volume of the wood increases, simultaneously lot of energy is liberated. This phenomenon is called imbibition. If such materials are kept in a closed container and water is added, wood imbibes water and swells. This swelling creates enormous pressure ranging from 1000 to 10,000 amphospheres. People (stone breakers) using this natural phenomenon break open big stone boulders by inserting dry wooden pegs into holes, then adding water in the holes. As wooden pegs imbibe water, enormous amount of imbibition pressure develops, which is really responsible for breaking the rocks.

Plants, being made up of cellulose cell walls, do exhibit imbibition. This is particularly conspicuous with xylem elements because their thick walls are made up of hydrophilic substances. When roots absorb water, xylem cells do imbibe water and there is no doubt about it; but the movement of water along the cell wall of these dead xylem water along the cell wall of these dead xylem elements seems to be incredible, but slow comparatively.

Instead of water moving along the cell wall it has been found that water moves through the lumen of the xylem elements and this is a fact. The contribution of imbibition force in the movement of water upwards is very little and negligible and it cannot be considered as the mechanism of ascent of sap.

COHESION / TRANSPIRATION PULL THEORY

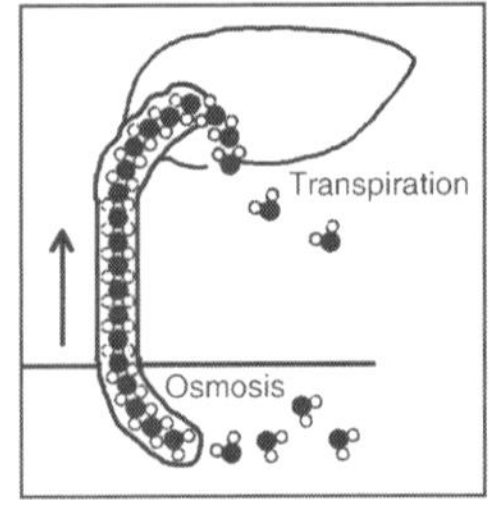

Fig. The cohesive force and the other is surface tension. The cohesive force that develops in such a column is enormous ranging from 100-1000 ATM.

Molecules of similar kind are attracted towards each other and they are held together by a force of attraction called cohesion force. Water, oil, alcohol etc., depending upon the kind of the solvent molecules the force of attraction varies. In the case of water, hydrogen and ionic bonds are the forces of attraction, while in oils the hydrophobic bonds are the forces of attraction.

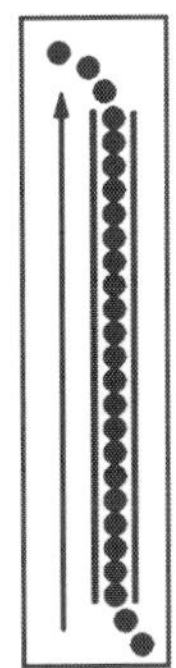

Fig. Plants Contain a Series of such Xylem Vessels Connected to one Another Forming a Continuous Column of Cells, form Basal Part of the Roots to the Terminal Leaves at the Extreme Apex on the Stem.

When such solvents are put into very fine tube similar to the size of tracheae, the column of the solvent, here it is water, does not break, because of two forces acting up on such column of water, one is the cohesive force and the other is surface tension.

The water found in such long columns is tenaciously held to the walls of the vessels, at the same time the column withstands the opposite pulls by its cohesive forces. When water is transpired from the surface cells of the leaf, water is drawn from the neighbouring cells due to the development of DPD gradient between the transpiring surface cells and the inner cells.

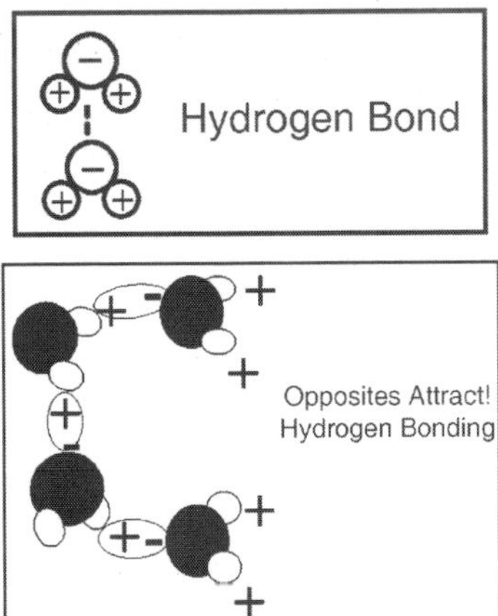

Fig. Hydrogen Bonding

This sets up a chain reaction, and xylem elements filled with water. This leads to the diffusion of water from xylem elements found in 'lumen into mesophyll cells towards the anterior surface. When millions of such cells are engaged in losing water by transpiration, enormous amount of DPD gradient develops which may amount for 100-200 atmospheres.

The totality of the force is so great on the water column found in xylem elements; the water is physically pulled upwards to meet the demand of the transpiring cells. This pull or force that drains the water is called transpiration pull or suction pressure. This kind of transpiration pull on the column upwards creates a kind of tension on the water column for the water column is also pulled downwards by gravitational force. Under these conditions, forces which have greater strength prevail and the water column is pulled towards that end. In this case, the transpiration pull is much more than the gravitation pull, hence the water is pulled upwards, provided that water column has to have strength or force of attraction to withstand tension. Fortunately the cohesive forces between the water molecules are so great, they can withstand this tension.

Thus water moves upwards in column as if there is a pump sucking the water upwards. Because of such tension the cells shrink, rather reduce their lumen, when the entire stem is taken into consideration. The stem exhibits a slight narrowing during maximum transpiration. These daily rhythmic movements of expansion (night) or contraction (day) has been demonstrated by Mc Dougal.

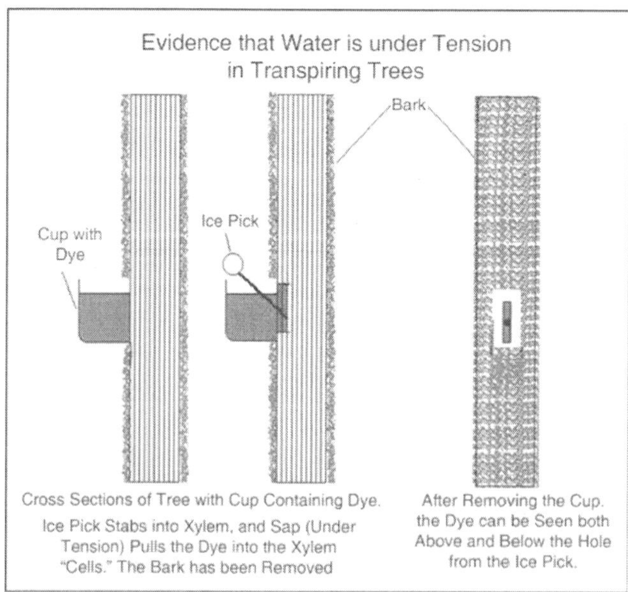

Fig. Cross Sections of tree with Cup Containing Dye.

One atmospheric pressure of transpiration pull is enough to pull the water unto the height of 20 feet or so; but normally, the transpiration pull that develops runs to about 20-100 atm, which is enough to pull the water to the height of 400 to 1000 ft and the tallest plant known to mankind is just about 400 ft. Certain objections like air bubble in water column are taken care of by this 'Cohesion theory by Dixon and Jolly (1894).

Thus it can be concluded that of all the physical phenomenon involved in ascent of sap, the cohesive-transpiration pull is the main mechanism responsible for the movement of water upwards.

That does not preclude the participation of other forces like imbibition pressure, capillary force, root pressure, but their contribution is very negligible. Still one should not forget that plant is living system. It uses the non-living structures like xylem elements for the transportation of water. Thus living cells around the dead cells are absolutely essential to keep this phenomenon going; without this no plant can either survive or performs the process. Hence, the vital forces also have a say in this process.

SOIL ACIDITY AND TOXICITIES

Elements essential for plant growth generally occur at soil concentrations that range between decient or adequate for plant growth. Instances do occur, however, where there are excess concentrations of either essential or non-essential elements. These toxicities may occur naturally, as saline soils, in soils developed from parent materials high in heavy metals, or through soil acidication. Human action can exacerbate natural processes. This has occurred with salinisation of the landscape as a result of clearing native vegetation or through irrigation. Heavy metals have accumulated in the landscape through industrial pollution, through the use of fungicides that contain Cu or the use of superphosphate which may contain Cd.

Accelerated acidication has occurred through certain farming practices and is a particularly serious problem on sandy soils with low buffer capacity. Application of N fertilizers and inclusion of pasture legumes in rotations increase acidication. Also, drainage of some estuarine soils for farming or urban uses, in which S is present in the reduced state (*e.g.* as iron pyrite) under waterlogged conditions (*i.e.* acid sulphate soils), results in oxidation of S and reduction in soil pH. Certain mine spoils are also acid, particularly those high in iron pyrite, which on oxidation result in acidication of spoil *in situ* or of run-off waters. Acid rain is a further problem, and in industrialised nations, especially in Europe and northeastern USA, there is considerable atmospheric pollution which produces acid rain containing dissolved oxides of N and S. This acid rain may fall long distances away from the source of pollution, and has led to international disputes.

SOIL ACIDICATION

As soils acidify, plant growth generally declines and there is a decrease in the range of plant species that may be grown. The activity of soil fauna and flora decreases also. The extent of acid soils and the degree to which plant growth is reduced by acid soil factors make these soils of considerable ecological and economic importance. Soil pH (negative logarithm of the molar activity of hydrogen ions) has been used to dene acid soils. Since pH is measured on a logarithmic scale, it is important to realise that a soil with pH 4 is 10 times more acid than a soil with pH 5! Acid soils may be classied as those soils with a pH of less than 7 (*i.e.* neutral pH). However, this has little value in practice, since most plant species will only show reduced growth at substantially lower pH. Thus, soils with a pH of less than 5.5 may be regarded as acid (*i.e.* those in which growth of many plant species is adversely affected).

Acid soils occupy about 30% of global land surfaces, and predominate in two major regions of the world: humid temperate forests and humid tropics and subtropics. However, many soils in other areas are acid also. Tropical acid soils alone comprise approximately two billion hectares, or 14%, of the total ice-free area of the world. Approximately 40% of the world's arable soils (*i.e.* those on which crops may be grown) are acid.

Eight regions of Australia have been identied in which agricultural production is reduced by acid soils. In southern Australia, these include the wheat and sheep belt of Western Australia, the southern portion of South Australia, cropping and grazing lands of southeastern South Australia along with the western, central and northeastern districts of Victoria. There are also about one million hectares of highly acidic soils in Tasmania. Further north in New South Wales, the cropping and grazing regions of the Riverina and southwestern slopes contain many acid soils, as do the tablelands and wet sub-tropical coastal farming regions of New South Wales and Queensland. Soils of the tropical wet coast of Queensland and adjacent intensively cropped and grazed areas are also acid to a considerable extent.

Overall, acid soils are most common where rainfall exceeds 450 mm year and more than 80 million hectares of the most productive agricultural land in Australia are acidic, with more than 40% of this land being highly acidic. Acid sulphate soils of coastal areas may also acidify when drained for crop production or urban use, with consequent effects on estuarine aquatic life. This most noticeably results in sh kills, often following rains after a dry period during which oxidation of soil S has occurred.

Table. Soil pH of Land in Southern Australia that has been used for Subternunean Dover pasture has shown a decrease in soil pH with time, as well as an increase in Exchangeable Al and Percentage of Soil Action-Exchange Capacity Occupied by Al (*i.e.* al Saturation). (Data Apply to the Surface 10can of Soil Profiles)

Pasture Age (cmol (+) kg soil)	Exchangeable Al (%)	Al Saturation	(Years)	Soil pH
	0–10	6.0	0.05	2
	15–30	5.4	0.28	10
	35-51	5.2	0.59	25

Acid soil infertility in Australia is now recognised as an insidious and invisible form of land degradation. Worldwide, soil nutrient decline and acidication are often the most easily visible forms of land degradation. A landmark Australian study showed that soils under subterranean clover pasture acidify, with a decrease from a soil pH of around 6.0 in virgin soils to pH 5.2 where pasture had been grown for more than 30 years. Subsequent studies have shown that the acid addition rate over extensive areas of New South Wales ranges from near zero to 3–5 kmol H ha year , with up to 20 kmol H ha year in some exploitative systems. With an acid addition rate of 4 kmol H ha year , there would be a decrease of about 1 pH unit in the surface 30 cm of a sandy loam within 30 years. This would take about 120 years in more strongly buffered clay soils.

ALUMINIUM AND MANGANESE TOXICITIES

Soil pH has a marked effect on nutrient availability by affecting both solubility of soil minerals present and the ability of plant roots to absorb nutrients. Generally, a decrease in soil pH increases availability of cations, especially essential micronutrients, Fe, Mn, Cu and Zn. In contrast, availability of Mo decreases, often to such an extent that acid soils are decient in plant-available Mo. Availability of P also decreases at low pH via xation with hydrous oxides of Fe and Al, although reactions with Ca can also reduce P availablility around pH 8.5.

Aluminium Toxicity

Al toxicity has been recognised for over 60 years. Shortly after the turn of the twentieth century, soil scientists found Al present in solution leached from soils. However, later attention was given to hydrogen clays, and following development of glass electrodes research emphasis was placed on soil pH for many years—an example of technology driving science! Despite publication of research results between the late 1920s and late 1960s, deleterious effects of Al on plant growth did not attract interest again until the early 1970s. Some of those ndings are outlined here.

Table. Many froms of Al are Potentially Present in Soil Solutions of Acid Soils (Not all are Toxic to Plants)

From	Examples
Micro-crystals	Kaolinite, Gibbsite
Amorphous Precipitates	Al_2SiO_5, $Al(OH)_3$, $AlPO_4$
Inorgaic Polycations	$Al_2(OH)_2$, $Al_3(OH)_3$ $AlO_4Al_{12}(OH)_{24}(H_2O)_{24}$
Organic Complexes	Al citrate, Al oxalate
	Al fulvate, Al humate
Inorganic Complexes	AlF , AlF_2 , $AlSO_4$
Inorganic Monomers	Al , AlOH , $Al(OH)_2$

Al is the most common metal in the earth's crust (8% of dry mass) and comprises some 7% of soils, where it has a complex chemistry. Al is an important component of aluminosilicate compounds, including clay particles. As soils acidify, Al dissolves from these solid forms, and enters solution with the potential to then become toxic.

Not all Al in the soil solution is toxic, however, providing challenges to biologists to identify those forms that are. Those shown to be toxic to plants include the inorganic Al monomers (Al , AlOH , $Al(OH)_2$) and the inorganic poly-cationAlO_4Al_{12} $(OH)_{24}$ $(H_2O)_{24}$ (known as Al_{13}).

The concentration of Al in the soil solution is often low (<50 μM), and activities of monomeric Al in the soil solution that reduce root growth have been found to range from 4 to 15 μM. A further challenge to analytical chemists is to develop techniques able to discriminate between toxic and non-toxic forms of Al, made all the more difcult because of the low concentration of toxic Al.

This concentration has been calculated to be about 0.1 g Al in 1 m of soil (*c.* 12 μM Al or one-third of one part per million in the soil solution). The same 1 m of acid soil would contain a total of 70 kg Al.

Solubility of Al decreases dramatically with an increase in soil pH, and the concentration of inorganic monomeric Al in solution is also affected by organic and inorganic anions. Complexes of Al with organic ligands of low relative molecular mass, especially citric and malic acids, are also considerably less toxic.

Tolerant plants benet from such chelating mechanisms. In addition, organic matter increases the concentration of fulvic and humic acids, resulting in the formation of Al fulvate and Al humate. These complexes are considerably less toxic than the inorganic monomeric species.

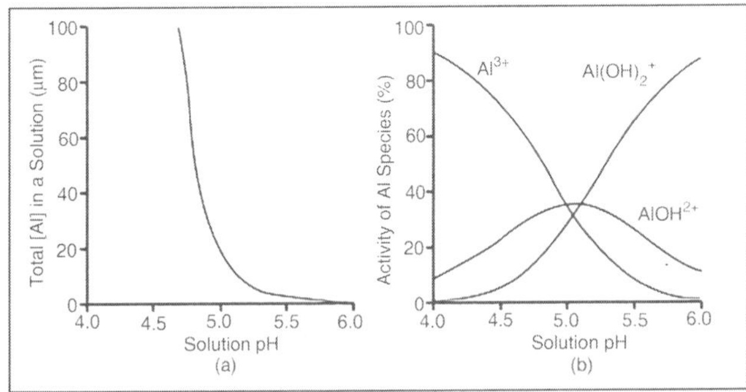

(a) Total concentration of all ionic species of Al in solution as a function of pH. In (b) a solution that initially contains 1000 μM $CaCl_2$ and 100 μM $AlCl_3$ (used to approximate a soil solution) shows a marked decrease in concentration of Al from pH 4.0 to pH 6.0, and a change in relative activities of different inorganic monomeric Al species.

Underlying equilibria responsible for such pH effects are summarised below. Values for pK coincide with the pH of a solution where the reaction mixture would be 50% dissociated. Increasing H+ ion concentration due to decreasing pH forces such equilibria towards Al , thus inducing Al toxicity.

$$Al + H_2O \Leftrightarrow AlOH + H \;\; pK = 5.00$$

$$Al + 2H_2O \Leftrightarrow AlOH + 2H \;\; pK = 10.10$$

Biological effects of ions are often better related to activ-ity in solution rather than to concentration. In addition to soil pH effects on Al solubility, speciation of Al in solution changes with pH. At pH 4.0, most of the inorganic monomeric Al is present as Al ; this decreases with an increase in pH while the activities of the hydroxy Al species (AlOH , $Al(OH)_2$) increase. At alkaline pH, aluminate ions (*e.g.* $Al(OH)_4$, $Al(OH)_5$) enter into solution but are not toxic to plant roots.

Soluble Al in solution has a rapid effect on root growth, with the effects often visible within 2 d. Microscopically, decreased root growth may be evident within a few hours of exposure to Al, and changes in Golgi apparatus activity have been documented to occur within 5 h of placing roots in a solution containing Al. To be toxic, the root tip must be exposed to Al.

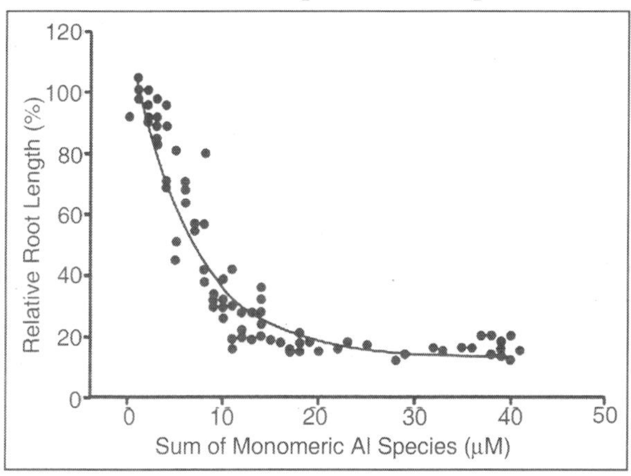

Fig. Soybean Root Growth Decreases Markedly as the Sum of Activities of Monomeric Al Species in Solution Increases.

While the toxic effect of Al has been known for more than 60 years, the biochemical basis of Al toxicity has not been claried. However, Al is known to exert its primary toxic effect on roots, initial effects including a reduction in root length (through reduced cell elongation) and damage to cortical cells of the root epidermis near the root tip.

Proliferation of root hairs, important for uptake of water, essential nutrients and for infection by N-xing rhizobia in legumes, is severely reduced by Al at concentrations lower than those that reduce root growth. Proposals for the primary toxic effect have included reactions of Al with components of the cell wall, plasma membranes, cytoplasm and nucleus.

Visible effects of soluble Al on plant tops are considered to be a secondary effect through reduced nutrient uptake. Symptoms include those similar to deciencies of Ca, Mg, Fe, and P, probably as a result of decreased root proliferation and of reduced root activity. Further secondary effects include a reduction in uptake of water, increased sensitivity to drought and decreased N_2 xation by nodulated legumes.

Plant species differ markedly in tolerance to soluble Al in the soil solution. As with toxicity, the biochemical basis of genetic tolerance of Al toxicity is not clearly understood. However, malic acid excretion by root apices of Al-tolerant wheat is known to be ve- to ten-fold greater than that excreted by Al-sensitive wheat.

Malic acid presumably complexes soluble Al, making it less toxic, and that capacity for excretion co-segregates with Al tolerance in progeny from crosses between near-isogenic lines. By implication, a single major gene is probably responsible for an Al tolerance that is functionally linked to Al stimulation of malic acid excretion.

Manganese Toxicity

Mn, an essential element for plants, is about the tenth most abundant element in the earth's crust. Like Al, Mn chemistry in acid soils is complex, affected by both soil pH and soil redox potential.

Mn occurs in rocks mostly in co-ordination with O and as the divalent Mn in soil solution and in natural waters. Biotic and abiotic oxidation occurs readily so that Mn does not generally occur to excess except in acid soils or in waterlogged soils (assuming there are sufcient Mn-containing minerals). Mn toxicity may occur following soil sterilisation due to loss of microorganisms that normally oxidise Mn.

Al is a root toxin, whereas high Mn is mainly toxic to shoots (at very high Mn concentration, root growth may be affected directly). Thus, unlike Al, which often has an immediate effect on root proliferation, Mn must accumulate in shoots to become toxic.

Three types of symptoms result from Mn accumulation: dark-brown, necrotic spots on lower leaves, distortion of expanding leaves (possibly an induced Ca deciency) and chlorosis of young leaves. Such chlorosis is often interpreted as an Fe deciency due to reduced Fe uptake by roots due to excess Mn.

Considerable genetic differences in tolerance to high concentrations of plant-available Mn exist among plant species, and even among lines within a species.

Such variation may occur via differences in root exclusion of excess Mn, complexation of Mn within roots, or shoot tolerance of high Mn. Sunflower, watermelon and cucumber all show shoot tolerance and even excrete Mn in a biologically inactive form around trichomes (hairs) on leaves and stems.

External Mn concentrations at which plant growth is reduced also vary greatly among plant species. In carefully controlled solution culture, the critical external concentration (*i.e.* the concentration required for 10% reduction in plant dry mass) in the two most sensitive species, maize and wheat, was 1.4 µM Mn. In contrast, sunflower growth was only reduced with 65 µM Mn in solution.

Table. Critical Concentrations of External and Internal Mn for a Number of Plant Species Grown in Solution Culture (*i.e.* Mn Concentrations Corresponding to 90% Maximum Dry Mass). Higher Concentrations would Inhibit Growth

Critical External Species	Critical Internal [Mn] (µm)	Mn retained in [Mn] (mg kg)	Plant Root (%)*
Maize	1.4	200	49
Wheat	1.4	280	59
Cowpea	1.9	720	5
Bean	2.2	280	11
Cotton	4.1	750	5
Soybean	9	600	33
Sweet potato	18	1380	7
Pigeonpea	46	300	32
Sunflower	65	5300	4

Likewise, the critical internal Mn concentrations varied also, from 200 mg kg in maize to 5300 mg kg in sunflower. Although cowpea and bean had a similar external critical concentration, cowpea was able to tolerate a much higher tissue Mn. A higher retention of Mn in roots of soybean enabled this species to tolerate a higher external Mn concentration than cotton.

Countering Adverse Effects of Acid Soils

Acid soil problems have been traditionally corrected via application of agricultural lime ($CaCO_3$) or dolomite ($CaCO_3$ + $MgCO_3$). Such amelioration raises soil pH, reducing Al and Mn concentrations in the soil solution, as well as adding the essential nutrients Ca and Mg. Lime or dolomite is generally required to raise soil pH to a level at which Al or Mn toxic-ities no longer affect plant growth (often above pH 5.5). Rates required vary considerably depending on soil type, with clay soils requiring higher rates to increase soil pH (greater buffering capacity referred to earlier in connection with acidication). Moreover, lime requirements for sustainable agriculture vary with farming systems.

For example, lime required (kg $CaCO_3$ ha year) to balance net acid accumulation is about: 0.8 for wool, 6.0 for lamb and 7–20 for cereal enterprises. Even though cereal production is not greatly acidifying, Australia's average wheat crop of 15 million t would have removed alkalinity equivalent to 135000 t $CaCO_3$ as well as 6000 t Ca, 18 000 t Mg and 66 000 t K (along with 330 000 t N, 42 000 t P and 26 000 t S) from the land. Most of the crop was used for human consumption either in Australian cities or overseas. There would be little return of these nutrients to their original elds.

Application of lime in Australia falls far below that required to balance losses of alkalinity. Annually, *c.* 0.5 million t of lime is used in Australia, but more like 2.25 million t are needed annually as prophylactic dressings. Overliming is a potential problem, especially on sandy soils, due to reduced availability of essential micronutrients, especially Zn, while root diseases such as 'Take all' of cereals are exacerbated.

Amelioration of some acid soils is possible via application of gypsum ($CaSO_4.2H_2O$) which provides a readily soluble form of Ca. Leaching soluble Ca applied as gypsum may improve root growth in acid subsoil horizons. Application of Mo is also benecial where this essential micronutrient is in short supply due to low pH.

Finally, breeding and selection for crop tolerance to acid soil limitations has immediate application in agriculture, but an understanding of factors involved is crucial because tolerance to one factor limiting growth on acid soils (Al toxicity) does not imply tolerance to another such as Mn excess or Mo deciency.

DIVERSITY AND ECOLOGY OF MYCORRHIZAL PLANTS

Mycorrhizas (sometimes called mycorrhizae or mycorrhizal associations) are by far the most common mutualistic symbioses involving vascular plants, and about 90% of higher plant families (both gymnosperms and angiosperms) contain species that can form mycorrhizal associations. Roots of most ferns also form mycorrhizas and similar structures occur in the absorbing organs of many bryophytes. Fossil mycorrhizas occur in rocks formed in the Devonian era, about 400 million years ago, and present-day mycorrhizas are the result of a very long period of worldwide co-evolution between plants and soil fungi.

Mycorrhizas fall into distinct classes, depending on the types of fungi and plant, and on the anatomy of the mycorrhizas themselves. Each main class shows wide specificity where a given species of mycorrhizal fungus can colonise different host species and one plant species can be colonised by different fungal species.

This situation contrasts sharply with the extremely narrow specicity of N_2-xing symbioses. There are broad differences between the dominant classes of mycorrhizas that are found in different ecosystems, depending on vegetation type, soil characteristics and growth-limiting nutrients. A walk up a mountain can pass through different mycorrhizal 'zones'.

Ectomycorrhizas (with their distinct external fungal sheath and with fungi penetrating intercellular spaces of the root cortex) occur on forest trees including eucalypts, beech (*Fagus* and *Nothofagus*), birch, oak and most conifers such as pine and spruce. Ericoid mycorrhizas (where fungi penetrate root cells) occur in plants of heathlands. They are also common in plants growing on peaty soils (those containing large quantities of organic matter) as on the Siberian tundra.

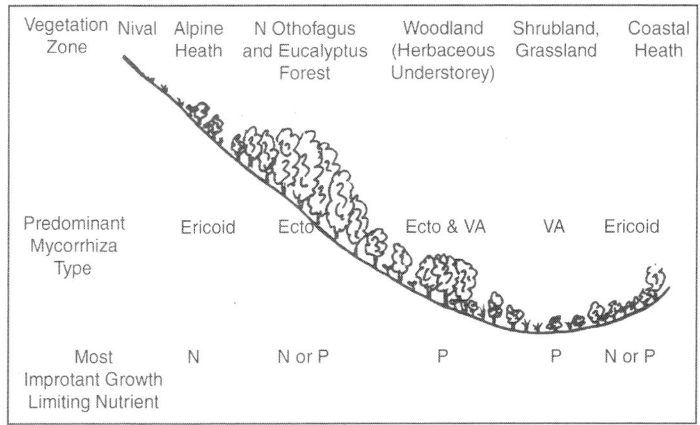

Fig. Representative Vertical Distribution of Mycorrhizal Classes on a Mountain, showing Major Types of Vegetation and Important Growth-limiting Nzutrients.

Overall, the most common mycorrhizal associations are vesicular-arbuscular (VA) mycorrhizas, where fungi again penetrate root cells and form nely branched arbuscules within those cells. Fungal storage bodies (vesicles) can form as either intercellular or intracellular structures. VA mycorrhizas predominate in understorey plants of woods and forests of all types, in temperate and tropical grasslands, in plants of scrub and desert and in many tropical forests.

Some classes of mycorrhizal fungi can grow in soil in the absence of a host plant and in pure culture, but VA mycorrhizal fungi cannot. A vast but unsuccessful research effort has gone into attempts to culture these fungi. Instant fame will be accorded whoever discovers the right medium!

The broad ecological zonations for different classes of mycorrhizas should not be overemphasised because the different classes can often be found very close together. For example, many trees of tropical rainforests form VA mycorrhizas but there are exceptions. Dipterocarps, which dominate many tropical rainforests, form ectomycorrhizas, while the other plants in the same forest will nearly all have VA mycorrhizas. Mediterranean-type vegetation—which is quite common worldwide—contains very mixed mycorrhizal associations, including ericoid, ecto- and VA mycorrhizas. Only in very inhospitable soils (*e.g.* saline, anaerobic and highly disturbed soils and those in very cold environments) are plants mostly non-mycorrhizal.

Since mycorrhizal symbioses are so common, research into the growth and nutrition of plants in the eld should always take into account their probable occurrence as a factor that can influence plant growth and competition, especially in soils where levels of essential inorganic nutrients are low. For example, research done in Western Australia has shown that disturbance of soils during surface mining can reduce the levels of mycorrhizal fungi, but survival in stored topsoil is still sufcient to help re-establish native vegetation when the topsoil is replaced after mining operations cease. Mycorrhizas can also improve productivity in cultivated plants even when levels of applied fertilizers are high. An important example in parts of Australia is the 'long fallow disorder' that reduces the growth of a wide range of crops when soils are left fallow for long periods. This effect with sorghum and sunflower grown in Queensland is due to a massive decline in populations of VA mycorrhizal fungi. Better growth of both plants on soils left fallow for shorter periods was associated with greater colonisation of roots by VA mycorrhizal fungi. There were higher levels of P per plant grown after short fallow and in sunflower P was higher even when expressed on the basis of% of dry weight; these effects reflected uptake of soil P via mycorrhizal fungi.

Table. Comparison of Healthy Crops Grown after Short Fallow Periods with Poor Crops Grown after Longer Fallow Periods (VAM = Vesiculararbuscular Mycorrhiz as)

Sorghum (8 Weeks Old)	Sunflower (16 Weeks Old)	Short Fallow	Long Fallow	Short Fallow	Long Fallow
Length of fallow	6	14	5	11	(months)
Dry mass (g plant)	58 ± 9	12 ± 2	152 ± 17	29 ±8	
% root length with	32 ± 2	3 ± 1	50 ± 6	8 ± 1	
P per plant (mg)	109 ± 27	18 ± 3	412 ± 53	65 ± 18	
P (as% of dry mass)	0.18 ± 0.02	0.16 ± 0.02	0.27 ± 0.01	0.22 ± 0.01	

Improved growth of forest trees after inoculation with specic ectomycorrhizal fungi has been demonstrated in a wide range of sites, including eucalyptus forests in Australia and elsewhere. These examples conrm an extensive interest in establishing and managing populations of mycorrhizal fungi in soils to increase plant productivity in agricultural and forest ecosystems.

BENETS AND COSTS OF SYMBIOSES

Physiological benets to a mycorrhizal plant include improved absorption of nutrients such as P and Zn via external fungal hyphae. Plant–water relations might also be improved, which would be important in arid conditions, but this is still a controversial area of research. Mycorrhizal fungi do not x atmospheric N_2 but can facilitate uptake of both inorganic and organic N and can broaden the range of sources of soil N that are available to plants. As shown in Queensland, mycorrhizal seedlings of *Eucalyptus* species can utilise sources of organic N that cannot be assimilated by non-mycorrhizal seedlings. In addition, external hyphae of mycorrhizal fungi can link plants of the same or (within limits) different species and can allow transfer of inorganic and organic compounds between plants. The idea that mycorrhizal plants can form linked and mutually benecial communities is an attractive one but it is not yet established if the amounts of solutes transferred from donor to recipient plants are always physiologically important. Transfer of organic C between linked plants is greatest when the recipient is shaded so that its photosynthesis is reduced. Accordingly, transfer of carbohydrate from parent plants might well assist seedlings to grow in deep shade as on a forest floor before they become exposed to bright sunlight.

Mycorrhizas in heterotrophic angiosperms such as some orchids lacking chlorophyll are physiologically distinctive. Even when mature, the plant drains the fungus of organic C derived from other sources (such as soil or donor plants that are autotrophic). This type of mycorrhizal fungus can also supply inorganic nutrients to the plant, but whether the fungi gain any nutritional benet remains to be established, so that such symbioses may not be truly mutualistic.

Where mycorrhizal fungi colonise autotrophic (photosynthetic) plants the fungus derives its organic C from the plant. This is a physiological cost to the plant but one which is usually not excessive, since mycorrhizal plants often grow faster than non-mycorrhizal plants of the same species under the same conditions. These positive growth effects are largest in shoots. Mycorrhizal plants sometimes grow root systems no larger or even smaller than those in non-mycorrhizal plants, diverting organic C to the fungus. However, there are some physiological conditions in which total growth of mycorrhizal plants is the same as or even slower than that of non-mycorrhizal plants of the same type. One of these conditions is low light intensity and here the cause of the effect is obvious: low light means that photoassimilate may be in short supply and the drain to the fungus may be deleterious to the plant.

The second condition resulting in slower growth of mycorrhizal plants is where levels of available soil P are high. The simplest explanation here is that the supply of soil P is adequate even for non-mycorrhizal plants and the C drain to the fungus (where present) then becomes a limitation on plant growth. Under these last conditions some plants reduce or prevent formation of mycorrhizas in a way that is analogous to the absence of N_2 xation where legumes grow in soils high in nitrate or ammonium.

Finally, some mycorrhizal fungi even 'cheat' their hosts by obtaining organic C without supplying P or other inorganic nutrients to their host. Why a plant tolerates being cheated in this way is not known. Lack of positive growth responses to colonisation is also sometimes found under a variety of eld conditions which are less easy to explain and are the subject of much research.

'Mycorrhizal responsiveness' summarises the most important factors which influence the extent to which a plant benets from the symbiosis by improved nutrition and growth. These factors include developmental and anatomical properties of the fungi, including their growth rates in soil, their ability to colonise plants and subsequent growth within roots, and their efciency in absorbing nutrients from soil and transferring them to the plant. Plants such as some cereals which have long, highly branched roots with long root hairs often show small mycorrhizal responses because the roots themselves are well structured for exploring large volumes of soil. Here, mycorrhizal hyphae confer relatively little extra advantage.

Table. Factor that can Influence Mycorrhizal Responsiveness, Excluding Mycorrhizas of Heterotrophic (Achlorophyllous) Plants

Fungus	Plant	Symbiosis	Community
External hyphae	Roots	Interfaces	
Growth rate	Growth rate	Area of contact	Interplant
	Connections	Extension into soil	Length
Carbon transfer	Plant density	(*i.e.* competition)	Colonisation rate
Branching	from roots		Nutrient uptake
Thickness	Nutrient transfer	Decreased	Nutrient translocation
Root hairs	to roots	infection by	
		pathogens	
	Nutrient uptake	Intraradicle hyphae	Carbon delivery to
	growth rate	interface	Nutrient delivery to
			interface

The interface between fungus and plant is obviously crucial in determining mycorrhizal responsiveness, a principle that applies equally to both intercellular and intracellular inter-faces. If there is a sufciently large area of contact between the symbionts, as in a VA mycorrhiza with many arbuscules, then there is the capacity for operation of a large number of transport proteins that catalyse selective transfer of nutrients. However, the rates of transfer are subject to regulation that can determine the relative amounts of C transferred to the fungus per amounts of P (or other essential nutrients) transferred to the plant and hence the nutritional costs and benets to the partners.

Plants were grown for eight weeks under controlled conditions in pots containing 0.5 kg of soil. The upper diagram (a) shows that growth stimulation of individual plants by VA mycorrhizas is greatest at low plant density. The lower diagram (b) shows that as the density of plants increases, total biomass per pot becomes more similar in mycorrhizal and non-mycorrhizal treatments due to limitations imposed by soil nutrient supply. Table also shows effects within plant communities that influence mycorrhizal responsiveness. Positive mycorrhizal growth responses decrease as the density of plants increases and competition for soil nutrients intensies. Put another way, where large numbers of roots and mycorrhizal fungal hyphae are close together, mycorrhizal roots appear to compete with each other and with any non-mycorrhizal roots, so that growth of all the individual plants is reduced compared to growth of plants spaced well apart. This effect is shown in Figure and may be one reason why intensive mycorrhizal colonisation fails to benet eld growth.

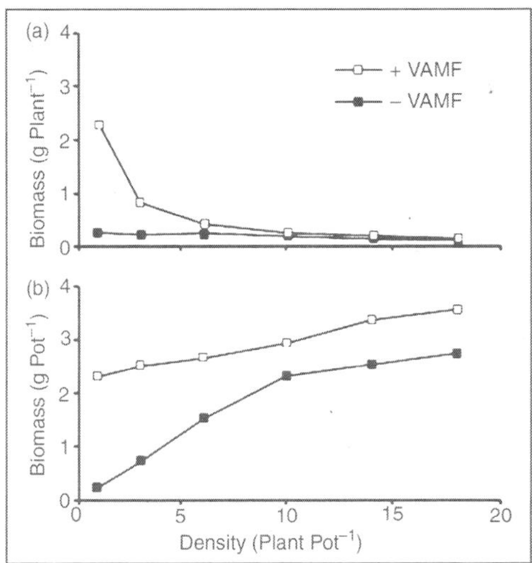

Fig. Effects of Plant Density on Growth of Mycorrhizal and Non-mycorrhizal *Trifolium Subterraneum*.

Hyphal links between plants of the same or different species complicates the issue of competition enormously. If nutrients or organic C are transferred via the hyphae between plants growing under similar conditions, what determines whether a plant is a donor or recipient of resources? This important issue awaits resolution.

Finally, mycorrhizal fungi can suppress infection by some pathogenic organisms—an obvious extra benet of a symbiotic association. Not surprisingly, responses of eld plants are more complicated and difcult to understand than in experiments with potted plants under controlled conditions!

SELECTION PRESSURES RESULTING FROM MYCORRHIZAL SYMBIOSES

A successful mycorrhizal symbiosis can act as a selection pressure. Such symbioses can affect growth and reproduction compared to non-mycorrhizal plants under equivalent conditions. This selection pressure can act within a population of a single plant species if the mycorrhizal fungus is present irregularly in the soil. Consider, for example, the *potentially* faster growth of an individual plant that becomes mycorrhizal compared to one that cannot or can do so only more slowly. The selection pressure can also act within a community of mixed species. Compare here the *potentially* faster growth of individuals of a species that forms mycorrhizas compared to individuals of a species that cannot. The subtle and complex nature of the selection pressures arises from the various features that can determine mycorrhizal responsiveness, which is why 'potentially' was emphasised. Nutritional benets in VA mycorrhizal plants can extend from increased vegetative growth to reproductive effects including earlier flower production, increased production of flower buds and increased seed production. In addition to improving plant fecundity, formation of mycorrhizas can improve the quality of plant offspring, for example their weight and nutrient content, and even the nutrient content of their seeds. These persistent effects are good evidence that formation of mycorrhizas can bring about strong effects on plant populations and on the structure of plant communities.

'NON-MYCORRHIZAL' SPECIES

Some terrestrial plants never form mycorrhizas even under growth conditions where colonisation would be expected. These plants belong to diverse families that are not closely related but include many chenopods and most brassicas (unfortunately including *Arabidopsis*). There are also individual 'non-mycorrhizal' genera within families that otherwise form mycorrhizas. One good example is the legume *Lupinus* (lupin). In such cases, an absence of these symbioses cannot be construed as disadvantage in evolutionary terms, so that mycorrhizal associations are not necessarily a universal advantage to terrestrial plants.

A different range of adaptive features allows these plants to grow and reproduce successfully in the absence of mycorrhizas. These include excretion of organic acids that aid solubilisation of P and production of long, relatively ne roots with long root hairs, which exploit soil P and other immobile nutrients in much the same way as do the external hyphae of mycorrhizal fungi, where present.

Notwithstanding alternative adaptations, an important question remains. What prevents colonisation by mycorrhizal fungi in non-mycorrhizal species? The answer is probably not the same for the different families and is still not known for many plants. In some cases, including brassicas, there is production of chemical defences against colonisation by mycorrhizal fungi. This response has similarities to chemical defence reactions in plant roots against pathogenic organisms.

QUANTITATIVE REQUIREMENTS

For practical use in crop production, plant nutritionists need to know whether a plant's internal supply of a nutrient is optimal for growth and, if not, how best to make it so. Answers to these two questions characterise a plant's nutrient requirements.

Unfortunately, discussions about a plant's quantitative nutrient requirements frequently fail to differentiate between the two quite distinct entities of internal and external requirements. Such confusion can be avoided with appropriate terminology: 'internal requirement' is the minimal concentration of a nutrient required within plant tissues and organs to sustain optimal metabolism and growth; 'external requirement' is the amount of a nutrient which must be supplied to soil or other culture medium for a plant to meet the internal requirement. Estimates of internal nutrient requirements of actively metabolising leaves of dicotyledonous plants are minimal concentrations for maximal growth. Internal require-ments of roots for Mn and Mg could be expected to be lower than those for leaves, while those of N_2-xing nodules for Co, which is not required by leaves, is around 0.1 mg kg.

Internal nutrient requirements of grasses and other monocotyledons appear similar to those of dicotyledons, except for a halving of their Ca and B requirements. With these exceptions and differences in the extent to which plants can use Na to replace K, variation in internal nutrient requirements offers little scope for improving nutrient use efciency of metabolic processes by selection or genetic mani-pulation.

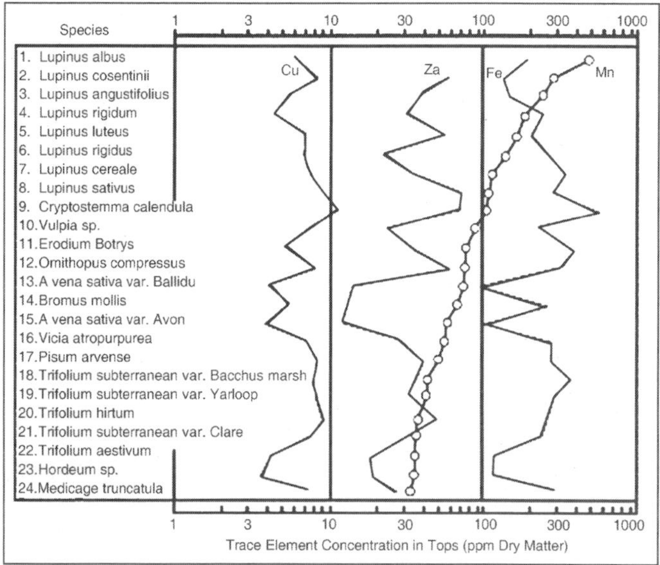

Fig. Trace Elements Represent only a Few Parts Per Million of Plant Shoot Tissue Dry Mass, but Nevertheless Vary between Species, and According to Varieties within Species.

By contrast, wide variation in shoot tissue concentrations of micronutrient ions acquired by different genotypes does occur and most notably for Mn. Such gene-based differences in acquisition of soil nutrients does constitute a ready source of variation in uptake efciency that can be put to good effect in breeding cultivars better adapted to decient soils.

These concentrations of Cu, Fe, Mn and Zn in tops of 24 annual crop and pasture species grown in filed plots at Gidgegannup, Western Australia, emphasise such variation and highlight an especially wide range of leaf Mn.

External requirements of seedlings may be modied by seed nutrient content, and in some cases seeds may contain sufcient Mo and Co for the life cycle of annual plants. Soil nutrient resources will vary in their availability to plants due to physico-chemical conditions in root zones, as well as fluctuations in root activity. Such intermittency in supply is buffered within plants by redistribution of internal nutrient resources as excess amounts in older tissues are retranslocated to growing regions.

Signicantly, not all nutrients can be recycled in this way, and in many plants Ca, B and sometimes Fe and Mn must be supplied continuously. Very little Ca or B moves from leaves to growing regions.

Indeed, roots cannot grow into environments lacking Ca or B, even when other parts of the same root system have ample supplies of these nutrients. Internal remobilisation of Fe and Mn embedded within plant tissues is similarly restricted.

This failure of Ca, B, Fe and Mn to move from leaves to growing regions has been attributed to their low phloem mobility. While this is probably true for Ca, restricted movement of B, Fe and Mn is better attributed more broadly to their immobility after deposition in leaves.

Immobile nutrients such as Ca contrast sharply with highly mobile nutrients such as N, P and K. Several other nutrients including S, Cu and Zn display variable mobility from leaves, being immobile from actively metabolising leaves but moving out of senescing leaves. B physiology remains an enigma with some species showing B mobility in association with sorbitol translocation. Those species translocating sucrose typically show low B mobility.

APPLICATION OF FERTILIZERS AND MANURES

Bulky organic manures should be applied well ahead of sowing, so that the preliminary decomposition takes place before the seeds germinate. Failing presowing application, they may be applied any time after the seedlings have established themselves.

They are best applied in the powedered form. A sufficient supply of moisture in the soil is essential for their rapid decomposition. In the case of inorganic fertilizers, potassic and phosphatic fertilizers are best applied just before sowing or transplanting. Nitrogenous fertilizers may be applied either at planting and partly later. Split application is particularly desirable for nitrogen when applied to irrigated crops or to crops in heavy-rainfall areas.

Fertilizers applied before sowing should be broadcast uniformly and harrowed in. In the case of fertilizers containing soluble phosphate, the desirability of applying them in 2.5 to 5 cm wide bands on each side of the row of seeds at a depth of 10 to 15 cm with a drill has already been pointed out.

This operation reduces the fixation of soluble phosphate in the soil. It is also a good practice to mix superphosphate with farmyard manure at 18 to 22 kg to a tonne before applying the organic manure, particularly of dairy farms land. Sulphate of ammonia used as a top-dressing should not be applied when plant leaves are wet. In the case of irrigated crops, the application of fertilizer should invariably be followed by a watering. In the case of fruit-trees, the fertilizer should be applied to the soil under the crown, a few metres away from the trunk. The area of application should be progressively extended as the trees grow bigger.

In advanced countries, fertilizers are usually applied with the help of machinery of diverse kinds and sometimes with an aeroplane or a helicopter. Combined-planters and fertilizer-distributors are employed when row crops are fertilized at sowing time.

DIAGNOSING THE FERTILIZER NEEDS OF SOILS

There are four methods of determining the fertilizer requirements of soil:

1. Field experiments,
2. Pot tests,
3. Biological tests,
4. Chemical test.

Field experiments contribute the more relaible method, but being time-consuming and expensive, they are conducted mainly by the research farms and research organisations. Farmers wishing to use field experiments as a Valid means of determining the fertility status of their soils should seek the advice of the state agronomist. Improperly conducted field experiments not only mean economic fertilizer practices.

Pot experiments permit test witha alarge number of manurial treatments within a limited space and in a relatively short time. However, as the conditions of such tests are different from those in the field, the results are not always directly applicable to large-scale farming.

Biological tests involve the growth of seedlings or of lower forms of plants, such as fungi and bacteria, under specified conditions and the study of their relative growth or the content of needed nutrients. A peridic testing of plant tissues for nitrates and other nutrients indicates the changing needs of crops for different food elements. But these are slow and costly processes and hence not always practicable.

The chemical analysis of soils or of plants growing on them constituents the modern method of determining the fertility status of a soil. Such analysis give information on the relative abundance or scarcity of the different nutrients required by crops from the soil, but they give no indication regarding the exact quantity, of fertilizer that may be applied to make good the deficiency. The dependability of this, method can, however, be increased a great deal by co-ordinating its results with those obtained from field experiments. Facilities for rapid soil-testing have been made available in almost all states and can be availed of. At the same time, a very large number of fertilizer experiments with different crops on a variety of soils are conducted annually in all parts of the country under a comprehensive scheme. The results of these field experiments, when calibrated against those of rapid soil tests, will make the latter purely dependable. This will then be a valuable aid in the hands of extension workers in furnishing advice to farmers regarding fertilizer practices.

It is also sometimes possible to obtain a clue to the nutrient deficiences of soil with the help of deficiency symptoms in plants, as described already. However, a correct deagnosis of deficiency symptoms needs extensive experience. Furthermore, such symptoms in plants appear long after the actual occurance of nutrient deficiency in the soil. Therefore, such soil deficiencies must be diagnosed and remedied much earlier by adopting other means.

How much Fertilizer to Apply

In a vast country, such as India, possessing a wide variety of climatic and soil conditions, no definite quantity of any fertilizer can be prescribed as an optimum dose even for one and the same crop for all regions. each state in the country has conducted fertilizer investigations for the last 50 years and accumulated information on the manurial requirements of principal crops under local conditions. Therefore, for specific fertilizer practices in relation to particular crops and soils, the state agricultural department should always be consulted. For determining the quantities of fertilizers from the recommended rates of application N,P or K, or *vice versa*, the following conversion factors may be used:

Table. Conversion Factors Determining Quantities of Fertilizers

Qunatity	Multiplied	Gives Corresponding by Quantity of
Nitrogen	4.854	Ammonium sulphate
Nitrogen	2.222	Urea
Nitrogen	3.846	Ammonium sulphate nitrate
Nitrogen	4.000	Ammonium chloride
Nitrogen	3.030	Ammonium nitrate
Phosphoric acid(P_2O_5)	6.250	Superphosphate, single
Phosphoric acid(P_2O_5)	12.222	Superphosphate, double
Phosphoric acid(P_2O_5)	2.857	Dicalcium phosphate
Phosphoric acid(P_2O_5)	5.000	Bone meal, raw
Potash (K_2O)	1.666	Muriate of potash
Potash (K_2O)	2.000	Sulphate of potash
Ammonium sulphate	.206	Nitrogen
Sodium nitrate	0.155	Nitrogen
Urea	0.450	Nitrogen
Ammonium sulphate nitrate	0.260	Nitrogen
Ammonium chloride	0.250	Nitrogen
Ammonium nitrate	0.330	Nitrogen
Superphosphate, double	0.450	Phosphoric acid(P_2O_5)
Dicalcium phosphate	0.350	Phosphoric acid(P_2O_5)
Bone meal, raw	0.200	Phosphoric acid(P_2O_5)
uriate of potash	0.600	Potash (K_2O)
sulphate of potash	0.500	Potash (K_2O)

Considerations Governing the use of Fertilizers

It may be stressed that in order to secure the maximum response to fertilizers, the crop should be irrigated immediately, following their application and at suitable intervals thereafter. The response of a crop under arid and semi-arid conditions is usually uncertain and relatively small. It is, therefore, advantageous to restrict the use of chemical fertilizers mainly ti irrigated lands and areas of assured rainfall.

It must also be borne in mind that the maximum profit from the use of fertilizers depends on many factors, such as the nature of soil, the kinds of crop grown, the climate (in relation to soil, the kinds of crops growth), the prices of fertilizers, the market price of the agricultural produce and so no. All these factors must be given due consideration to secure the most economic results. Changes in one or more of them are bound to influence the economic results. changes in one or more of them are bound to influence the economics of fertilizer use. The use of fertilizers is also subject to the 'law of diminishing returns'. This means that the rate of increase in the crop yield decreases after a certain point is reached regarding the quantity of fertilizer used, and consequently the value of the additional yield finally becomes less than the cost of the fertilizer.

Usually, the smaller applications of fertilizers produce greater percentage increase of yield than larger applications. It should also be exphasized that an adequate supply of nutrients is only one of the factors that determine crop yield and that the application of fertilizers is not the sole means of making good the nutrient deficiencies in soils and plants. Equal attention must be paid to the other soil and crop-management practices to ensure good tilt, proper drainage, the required soil reaction, soil conversion, good land use, a suitable crop rotation, adequate organic matter in the soil and satisfactory soil micro-organism activity.

Each one of these plays a vital role in determining the eventual productioin. the neglecting of one or more or these factors leads to a reduction in yield and created the need for still heavier manuring. Finally, manuring should not only be balanced in itself, but should also be designed to supplement the good effects of proper land use and beneficial soil management.

Table. Conversion Factors

Distance		
1 mile	=	1760 yards = 5,280 feet
1 metre	=	1.0936 yards = 39.37 inches
1 yard	=	0.9144 metre
1 foot	=	0.3048 metre
1 inch	=	2.54 centimetres

Area	
1 acre	= 4840 sq yards = 43,560 sq feet
1 *guntha*	= 1/40 acre = 1089 sq ft = 33' * 33'
1 acre	= 0.4047 hectare
1 hectare	= 2.471 acres
1 sq mile	= 640 acres

Weight		
1 long ton	=	2,240 pounds
1 short ton	=	2,000 pounds
1 metric ton	=	1000 kilograms = 2204.6 pounds
1 long ton	=	28 maunds(approx.)
1 1 maund	=	80 pounds(approx.)
1 metric ton	=	1.1023 short tons
1 metric ton	=	0.9842 long ton
1 short ton	=	0.9072 metric ton
1 long ton	=	1.0161 metric ton
1 pound(avoir)	=	453.6 grammes = 0.4536 kg
1 hundred-weight (cwt)	=	112 pounds = 50.8 kg
1 kilogramme	=	2.2046 pounds

Weight per area		
1 lb per acre	=	1.12 kilogrammes per hectare
1 cwt per acre	=	125.6 kilogrammes per hectare

Capacity		
1 gallon(Imp.)	=	4.5461 litres
1 gallon (U.S.)	=	3.7853 litres
1 gallon (Imp.)	=	0.1604 cft
1 litre	=	0.2200 gallon (Br.)
1 litre	=	2642 gallon(U.S.)
1 bushel	=	56 lb (potaoes)
1 bushel	=	56 lb (barley)
1 bushel	=	60 lb wheat
1 bushel	=	42 lb (oats)
1 gallon (Imp.) water weighs 10 lb		
1 cft of water weighs 62.3 lb		

PLANT NUTRIENTS AND THEIR FUNCTIONS

The plants require, the following essential nutrients for their normal development:

Carbon	Nitrogen	Calcium
Hydrogen	Phosphorous	Magnesium
Oxygen	Potassium	Sulphur
Iron	Zinc	Chlorine
Manganese	Boron	..
Copper	Molybdenum	..

Carbon is obtained from carbon dioxide of the air; Oxygen from air and water; Hydrogen from water; Nitrogen from air and soil or both, and all other nutrients from the soil. Soil is a the most important source of plant food.

Nitrogen, Phosphorous and Potassium are known as primary plant nutrients; Calcium, Magnesium and Sulphur are secondary nutrients; Iron, Manganes, Copper, Zinc, Boron, Molybdenum and Chlorine as trace elements or micronutrients. The primary nutrients and secondary nutrients elements are known as major elements. This classification is based on their relative abundance, and not their relative importance. the micronutrients are required in small quantities, but they are as important as the major elements in plant nutrients.

Air is the primary source of Nitrogen for plant nutrient. Only leguminous crops can directly use this free Nitrogen with the help of symbiotic bacteria of the genus *Rhizobium*. Other plant derive from soil their Nitrogen in the form of Nitrogen and Ammonium. Nitrogen and Ammonium are produced in the soil by action of micro-organisms on the soil organic matter. Non symbiotic micro-organisms can fix free Nitrogen of the air and make it available to plant in Ammonium and Nitrates forms. *Nitrogen* encourages the vegetative development of plants by importing a healthy green colour to the leaves. It also controls, to some extent the efficient utilization of phosphorous and Potassium. Its dependency retards growth and root development, turns the foilage yellowish or pale green, histens maturity, causes the shrivelling of grains and lowers crop yield. The older leaves are affected first. An excess of Nitrogen produces leathery(sometimes crinkled), dark-green leaves and succulent growth. It also delays the maturation of plants, impairs the quality of crops like barley, potato, tobacco, sugarcane, and fruits; increases susceptibility to diseases and causes 'lodging' of cereal crops by inducing an undue lengthoning of the stem internodes.

Phosphorous influences the vigour of plants and improves the quality of crops. It encourages the formation of new cells, promotes root growth(particularly the development of fibrous roots), and hastens leaf development through emergence of ears, the formation of grains, and the maturation of crops. It also increases resistence to diseases and strengthens the stems of cereal plants, thus reducing their tendency to lodge. It offsets the harmful effects of excess nitrogen in the plant. When applied to leguminous crops, it hastens and encourages the development of nitrogen-fixing nodule bacteria. If phosphorous is deficient in the soil, plants fail to make a quick start, do not develop a satisfactory root-system, remain stunted and sometimes develop a tendency to show a reddish or purplish discolouration of the stem and foilage owing to an abnormal increase in the sugar content and the formation of anthoscyanin.

However, the deficiency of this element is not easily recognised as that of nitrogen. It has also been observed that cattle feeding on the produce of deficient soils become dwarfed, develop stiff joints and lose the velvetty feel of the skin. Such animals show an abnormal craving for eating bones and even soil itself.

Potassium enhances the ability of plants to resist diseases, insect attacks, and cold and other adverse conditions. It plays an essential part in the formation of starch and in the production and translocation of sugars, and is thus of special value to carbohydrate-rich crops, *e.g.* sugarcane, potato and sugar-beet.

The increased production of starch and sugar in legumes fertilized with potash benefits the symbiotic bacteria and thus enhances the fixation of nitrogen. It also improves the quality of tobacco, citrus etc. With an adequate supply of potash, cereals produce plump grains and strong straws. But an excess of element tends to delay maturity, though, not to be the same extent as nitrogen.

Plants can make up and store potassium in much larger for correcting zinc deficiency. The symptoms of zinc deficiency appear generally in younger leaves, starting with intervienal chlorosis leading to a reduction in shoot growth and the shortening of internodes. Mottle leaf, little leaf, etc. in the case of trees are symptoms of zinc difficiency. The buds of several defficient maize plants become white; in citrous interveinal chlorosis and mottled leaf occur. In calcareous soils and in soils with very high phosphorus content, zinc defficiency is commonly expected to occur.

The principal function of zinc in plants is as a metal activator of enzymes. In highly weathered coarse textured soils zinc deficiency appears under an intensive cropping programme. The availability of zinc is least between pH 5.5 and 7, but its availability increases at a lower pH. At a higher pH above 7, zinc availability becomes a complex problem, as the positively charged zinc ion gets converted into a negatively charged zincate complex whose availability tends to be reduced in alkaline soils. When the calcium ion is predominent, the highly insoluble calcium zincate is formed and zinc availability gets seriously limited.

The application of soluble zinc salts or zinc chelates to the soil is generally recommended to correct its defficiency. Foliar sprays are advocated, especially for orchard trees for amending zinc defficiency. About 5 to 50 kg of zinc sulphate per hectare is used for such purposes.

The symptoms of *boron* defficiency vary with the kind and age of the plant, the conditions of growth and the severity of the deffiency. Each crop produces its characteristic growth abnormalities associated with boron defficiency, such as yellows and rosetting in lucerne, snakehead in wallnuts, die-back and corking of fruits in apple, corking and pitting of fruits in tomatoes, hollow stem and the bronzing of curd in cauliflower, the brown-heart diseases in table-beets, turnips, etc.

Molybdenum deficiency produces whip-tail in cauliflower, broccoli and other *Brassica* crops. The deficiency of this element reduces the activity of the symbiotic and non-symbiotic nitrogen-fixing micro-organisms. It was in 1954 that *chlorine* was proved to be an essential micronutrient. Its defficiency under field conditions has not been reported so far. In water-culture solutions, the leaves of chlorosis, necrosis and an unusual bronze discolouration on tomatoes.

Sodium is not an essential element for plant growth. But some crops, such as beet, celery, cabbage, kale, knol-khol, radish, rape and turnip, benefit greatly by application of soluble sodium salts, specially if the soil is deficient in potassium. Sodium is also of direct benefit to plants indigneous to the sea-shore or to irrigated arid regions. Salts of this element are said to release more of potassium from the exchange complex and to help to maintain phosphorus in a more available form. They also serve as a partial substitute for potassium in the case of potatoes and cotton.

MAINTENANCE OF SOIL FERTILITY

No two soils are alike either in respect of their nature or in respect of quantities of plant nutrients they contain. Under a given situation, the system of farming, soil management and manuring practices, etc., influence the productiovity of soils and crop yields obtained from them. The quantities of the three primary nutrients- N, P_2O_5 and K_2O removed from a hectare of land by some of the important crop are shown in table.

It is estimated that the different agricultural crops in India remove about 4.27 million tonnes of nitrogen, 2.13 million tonnes of phosphoric acid, 7.42 million tonnes of potash and 4.88 million tonnes of lime per year. The production of larger yields through improved varities of crops and intensive cultivation will increase the depletion of nutrients still further. But erosion and leaching cause additional losses. The present production of synthetic nitrogenous fertilizers in the country reached 1.5 million tonnes of nitrogen in 1975-76, and the bulky organic manures might supply another 1.5 million tonnes of total nitrogen. The amounts of phosphorous, potash, etc., added to the soil are very small.

9

Photosynthesis

Photosynthesis is the process by which plants, some bacteria, and some protistans use the energy from sunlight to produce sugar, which cellular respiration converts into ATP, the "fuel" used by all living things. The conversion of unusable sunlight energy into usable chemical energy, is associated with the actions of the green pigment chlorophyll. Most of the time, the photosynthetic process uses water and releases the oxygen that we absolutely must have to stay alive.

Photosynthesis is the process by which light energy is captured, converted and stored in simple sugar molecule. This process occurs in chloroplasts and other parts of green organisms. It is a backbone process, in the sense that all life on earth depends on it's functioning.

The following equation sums up the process:

$$6CO_2 + 12\ H_2O + \text{light energy} \rightarrow C_6H_{12}O_6 + 6O_2 + 6H_2O$$

Carbon water Glucose Oxygen Water

As you see from the equation, this process is vital to us as humans, because it transforms carbon dioxide into oxygen—which we enjoy with every breath!

CARBON DIOXIDE (CO_2)

The earth's atmosphere contains approximately 79% nitrogen, 20% oxygen and the remaining 1% is a mixture of less common gases—including 0.039% carbon dioxide. Carbon dioxide in the atmosphere reaches plant mesophyll via the stomata. The carbon dioxide dissolves on the thin film of water that covers the outside of cells. The carbon dioxide then diffuses through the cell wall into the cytoplasm in order to reach the chloroplasts. The oceans hold a large reservoir of carbon dioxide, which keeps the atmospheric levels essentially constant. Although there are some indicators that the atmospheric levels of CO_2 are rising and adding to the global warming issue. That is a whole other topic though.

WATER

Water is plentiful on earth, however, it may or may not be plentiful at the location of each individual plant. Therefore, plants will close their stomata, if need be, which reduces the CO_2 supply to the mesophyll. Not even 1% of the water that is absorbed by plants is used in photosynthesis, the remainder is either transpired or incorporated into protoplasm, vacuoles or other cell materials. The water utilized in photosynthesis is the source of oxygen released as a photosynthetic byproduct.

LIGHT

Light has a dual nature, in that it exhibits properties of both waves and particles. The energy from the sun comes to earth in various wavelengths, the longest being radio waves and the shortest are gamma rays. Approximately 40% of the radiant energy the earth receives from the sun is visible light. Visible light ranges from red, 780 nanometers to violet, 390 nanometers.

The violet to blue and red to orange ranges are the most often used in photosynthesis. Most light in the green range is reflected. Of the visible light that reaches a leaf, approximately 80% is absorbed. Light intensity varies widely. Time of day, temperature, season of year, altitude, latitude and other atmospheric conditions all play roles in the intensity of the radiant energy that will reach the earth and it's organisms. High intensity light isn't necessarily a beneficial thing for plants. In high intensity light, photorespiration may occur, which is a type of respiration that uses oxygen and releases carbon dioxide but differs from standard aerobic respiration in the pathways that it utilizes.

CHLOROPHYLL

A few things to know about chlorophyll before we get into the nitty gritty of photosynthesis and respiration. There are more than one type of chlorophyll, however, they all have one atom of magnesium in the center. In some ways the chlorophyll is quite analogous to the heme structure in hemoglobin (the iron containing pigment that carries oxygen in blood). Chlorophyll has a long lipid tail that anchors the molecule in the lipid layers of the thylakoid membranes—recall that thylakoids are coin-like discs in stacks within the stroma of the chloroplasts.

The chloroplasts of most plants contain two types of chlorophyll imbedded in the thylakoid membranes. The formula for bluish-green chlorophyll a is $C_{55}H_{72}MgN_4O_5$ and the formula for yellow-green chlorophyll b is $C_{55}H_{70}MgN_4O_6$. In general, most a chloroplast has about three times as much chlorophyll a than b. The main role of chlorophyll b is to broaden the spectrum of light available for photosynthesis: chlorophyll b absorbs light energy and transfers the energy to a chlorophyll a molecule. Other pigments are contained in chlorophyll c, d, and e and take the place for chlorophyll b in some cases. Note that all the chlorophyll molecules are related to each other and differ only slightly in molecular structure. Light-harvesting complexes contain 250 to 400 pigment molecules and are referred to as a photosynthetic unit. There are countless numbers of these units spread throughout the grana of a chloroplast. In the chloroplasts of green plants, two types of these harvesting units operate together in order to bring about the first phase of photosynthesis.

The photosynthetic process occurs in two successive processes: the light reactions and the carbon-fixing reactions.

The Light Reactions

The light reactions involve light striking the chlorophyll molecules embedded in the thylakoids of chloroplasts. The subsequent reaction results in the conversion of some light energy to chemical energy. In the light reactions, water molecules are split apart into hydrogen ions and electrons and oxygen gas is released. In addition, ATP (adenosine triphosphate) molecules are created and the hydrogen ions derived from the water molecules are involved in "loading" NADP which carries the hydrogen as NADPH. NADPH is integral in providing the hydrogen ions used in the second series of major photosynthetic reactions: the carbon-fixing reactions.

The Carbon-fixing Reactions

The carbon-fixing reactions used to be called the dark reactions because light does not play a direct role in their functioning. The reactions take place in series outside of the grana in the stroma of the chloroplast. These reactions only occur if the end products of the light reactions are available for use. Depending on the plant involved, the carbon-fixing reactions may develop or progress in different ways. The most common type of carbon-fixing reactions in plants is the process called the Calvin cycle. In the Calvin cycle, carbon dioxide from the atmosphere is combined with a 5-carbon sugar—RuBP, or ribulose bisphosphate. The combined molecules are converted via several steps into a 6-carbon sugar, such as glucose. The ATP and NADPH molecules from the light reactions provide the energy and resources for the reactions. Some of the sugars produced are further combined into polysaccharides (strings of simple sugars) or are stored as starch within the plant. There are other variations, including the 4-carbon pathway which is usually found in desert plants (C_4 plants).

Before getting into respiration let's take a closer look at what happens in both the light reactions and the carbon-fixing reactions.

NITTY-GRITTY OF LIGHT REACTIONS

Einstein called the discrete particles of light photons. Particles (photons) and waves are both currently accepted aspects of light. The quantum (energy) of photons is different depending on what kind of light they are in. Longer wavelength light has lower photon energies, while light with shorter wavelengths have higher photon energies. As mentioned earlier, every pigment colour has a different distinctive pattern of light absorption—called the pigment's absorption spectrum. The energy levels of some of the pigment's electrons are raised when the pigment absorbs light. If energy is emitted immediately upon absorption, the effect is called fluorescence. The red part of light does this characteristically, as demonstrated when chlorophyll is placed in light it will appear red. If the absorbed energy is emitted as light after a delay, then the effect is called phosphorescence. The energy may be converted to heat or stored, as in photosynthesis within chemical bonds.

OXIDATION-REDUCTION REACTIONS

OIL RIG, a cute little mnemonic device to remember that oxidation is loss and reduction is gain. Perhaps better put, oxidation results in the net loss of an electron or electrons, while reduction results in a net gain of an electron or electrons. The electrons come from compounds within the process or donated in from previous processes. These types of chemical reactions are found scattered throughout the processes within photosynthesis and respiration.

PHOTOSYSTEMS

The two types of photosynthetic units in most chloroplasts are what constitute photosystem I and photosystem II.

Photosystem

contains photosynthetic units with 200 or more molecules of chlorophyll a, small amounts of chlorophyll b, protein saddled carotenoid pigment, and a pair of specialized reaction-center molecules of chlorophyll called P_{700}. All pigments in a photosystem are capable of absorbing photons, however, only the reaction-center molecules can really utilize the light energy. The other pigments aren't worthless in the system, as they act sort of like an antenna in gathering and passing light energy along to the reaction-center. Iron-sulphur complexed proteins initially receive electrons from P_{700} and serve as primary electron acceptors for the unit.

Photosystem

Contains chlorophyll a, protein saddled beta-carotene, a small amount of chlorophyll b and special pair of reaction-center molecules of chlorophyll a otherwise called, P_{680}. The photosystem has a primary electron acceptor called pheophytin or Pheo. For the record, the 680 and 700 in the names of the reaction-center molecules stands for the peaks in the absorption spectra of light waves of 680 nm and 700 nm.

PHOTOLYSIS

A photon of light strikes the photosystem II reaction-center, the P_{680} molecule to be exact near the inner surface of a thylakoid membrane. The received light energy excites an electron (boosts it to a higher energy level) which is an unstable reaction and thus most of the energy is lost to heat. Up to four photons at a time can strike the P680 molecule, however, it can only accept one electron at a time. The molecule of pheophytin picks up the excited electron, which then crosses the thylakoid membrane and is passed along to another acceptor called plastoquinone or Pq near the outside surface of the thylakoid membrane. Protein Z extracts electrons from water and replaces the ones lost by the P680 molecule. Protein Z contains manganese which is required in order to split water molecules. Simultaneously, as two water molecules are split and molecule of oxygen and four protons are produced. This enzyme-mediated water splitting process is called photolysis.

OVERALL EQUATION OF PHOTOSYNTHESIS

Photosynthesis is the intracellular anabolic process, characteristic of the green cells of plants in which carbohydrates are synthesized from carbon dioxide and water in the presence of light and chlorophyll. In this process, light energy is converted into chemical energy and stored in carbohydrate molecules, while oxygen is liberated.

The overall process of photosynthesis is represented by the following general chemical equation.

$$6CO_2 + 12H_2O \xrightarrow[\text{Chlorophyll-a}]{\text{Light}} C_6H_{12}O_6 + 6H_2O + 6O_2$$

PRIMARY PROCESSES OF PHOTOSYNTHESIS

Nature of Light

As stated earlier, solar radiation is the only natural source of light for all organisms. Green plants can utilize light belonging to the visible spectrum only for photosynthesis. This visible part of the electromagnetic spectrum from the sun is confined between wavelengths of 390 nm (violet) and 760 nm (red).

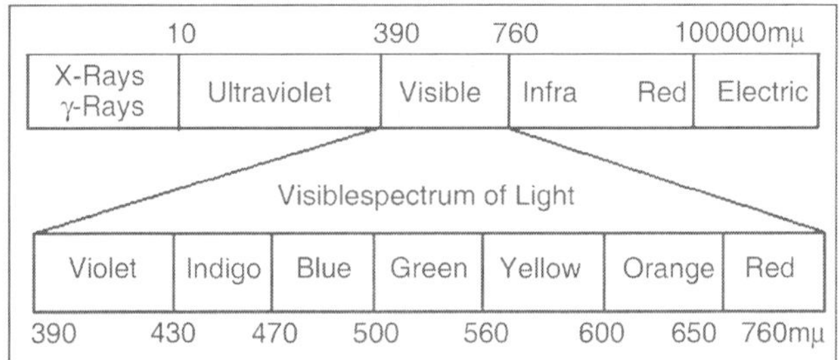

Light is a form of energy. It appears to travel as a stream of discrete particles called photons. Each photon contains one quantum (unit) of light energy. The energy quantity of each type of light depends on its wavelength. Light with a shorter wavelength has greater energy; light with a longer wavelength has lesser energy.

Rate of Photosynthesis

Rate of photosynthesis is measured in terms of the amount of carbon dioxide used (reduced) or oxygen released. It has been estimated that 8 quanta (photons) of light are required to reduce each molecule of carbon dioxide (or to release each molecule of oxygen) during photosynthesis (Emerson and Lewis, 1943). This is called the quantum requirement in photosynthesis.

Light Trapping Systems

During the photochemical phase, light is trapped by the photosynthetic pigments present in the quantasomes of the grana thylakoids. These pigments are organized into two pigment systems called pigment system I (PS I) and pigment system II (PS II).

(I) PS I is composed of the following pigment molecules:

Chl-a 700 (P700)	– one molecule	– Reaction center
Chl-a 683	– 200 molecules	– Antenna chlorophyll
Carotenoids	– 50 molecules	– Accessory pigments

(II) PS II is composed of the following pigment molecules:

Chl- a 680	– one molecule	– Reaction center
Chl- a 670	– 200 molecules	– Antenna chlorophyll
Chl- b	– up to 200 molecules	} Accessory pigments
Carotenoids	– up to 50 molecules	

In both the systems, all pigment molecules help in trapping light energy. However, all other molecules transfer their energy to the reaction center to be used for the photochemical reactions in photosynthesis.

Photo-excitation of Chlorophyll-a

When a molecule of chlorophyll-a (acting as the reaction center) receives light energy in the form of photons (quanta), it becomes energy-rich and activated. This is called the excited state of chlorophyll. In this excited state, the chlorophyll-a expels one electron. With the loss of an electron, the chlorophyll-a develops a positive charge. This is called ionized chlorophyll-a.

The expelled electron contains an extra amount of energy received from light. In other words, the energy-rich expelled electron represents light energy. This electron energy is used for the formation of ATP during photosynthesis.

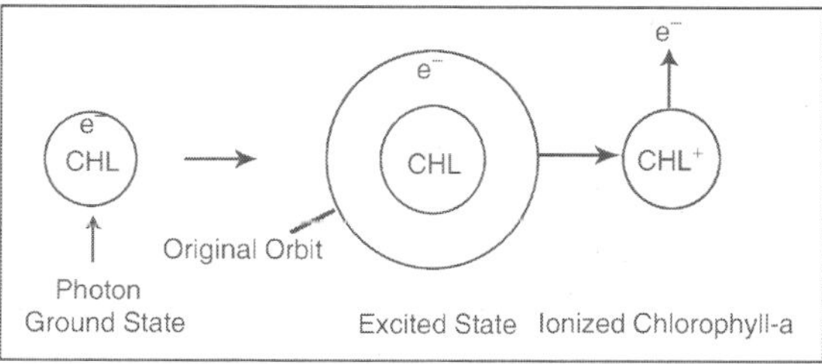

Photolysis of Water

In the photosynthesis of green plants, water is used as a source of hydrogen required for the reduction of carbon dioxide to form carbohydrates.

$$CO_2 + 2H_2O \xrightarrow[\text{Chlorophyll}]{\text{light}} \underset{\text{carbohydrate}}{(CH_2O)} + H_2O + O_2$$

In this process, water is oxidized in presence of light and chlorophyll. Hydrogen is removed from water, and oxygen is released as follows:

$$4H_2O \xrightarrow[\text{Chlorophyll}]{\text{light}} 2H_2O + 4H^+ + 4e^- + O_2$$

$$\text{or } 2H_2O \rightarrow 4H^+ + 4e^- + O_2$$

This is described as photolysis of water, or photochemical oxidation of water. This was first indicated by Van Niel (1931). It was demonstrated with the help of an experiment by R. Hill (1937). Hill suggested that during the photolysis of water, hydrogen combines with some hydrogen acceptor in the plant, while oxygen is released.

$$2H_2O + 2A \xrightarrow[\text{Chlorophyll}]{\text{light}} 2AH_2 + O_2$$

(where A = unknown H-acceptor in plants).

Later on, Arnon (1951) discovered this hydrogen acceptor to be a coenzyme called NADP (nicotinamide adenine dinucleotide phosphate).

Substance A = NADP

$$2H_2O + 2NADP \xrightarrow{\text{light}} 2NADPH_2 + O_2$$

Chlorophyll reduced co–enzyme

Similarly, Ruben, Kamen, *et al.* (1941) confirmed that the oxygen evolved during photosynthesis comes from the splitting of water only. They used heavy isotope of oxygen (O_{18}) to demonstrate the fact as follows:

$$\text{(I) } 6CO^{18}{}_2 + 12H_2O \xrightarrow[\text{Chlorophyll}]{\text{Light}} C6H_{12}O^{18}{}_6 + 6H_2O^{18} + 6O_2$$

$$\text{(II) } 6CO_2 + 12H_2O^{18} \xrightarrow[\text{Chlorophyll}]{\text{Light}} C6H_{12}O_6 + 6H_2O + 6O^{18}{}_2$$

HILL REACTION

R. Hill (1937) suspended isolated chloroplasts in water in a beaker. He added some known hydrogen accepting compound (*e.g.* benzoquinone or ferric salt) to the water and exposed the beaker to the light. Immediately the oxidation of water took place and bubbles of oxygen were evolved. When the hydrogen acceptor in the beaker was examined, it was found to be reduced.

$$2H_2O + 2A \xrightarrow[\text{Chlorophyll}]{\text{Light}} 2AH_2 + O_2$$

(A = hydrogen acceptor)

This is photolysis or the photochemical oxidation of water and is commonly called the Hill reaction.

Photophosphorylation

The formation of ATP molecules (chemical energy) from ADP and H_3PO_4 in presence of light and chlorophyll-a during the photochemical phase of photosynthesis is called photophosphorylation or photosynthetic phosphorylation. More commonly, it is described as the conversion of light energy into chemical energy (ATP).

$$ADP + H_2PO_4 \xrightarrow[\text{Chlorophyll-a}]{\text{Light}} ATP$$

(ADP = adenosine diphosphate)

During the light reactions (the primary process or the photochemical phase), ATP formation takes place through two types of phosphorylation reactions.

These are:

The non-cyclic process involves both PS I and PS II. The cyclic process involves only PS I.

- Non-cyclic photophosphorylation

In the light reaction of photosynthesis, the formation of ATP from ADP and H_3PO_4 in presence of light and chlorophyll-a during the non-cyclic transfer of electrons is called non-cyclic photophosphorylation.

Important Features:

- It is the major pathway of the light reaction of photosynthesis.
- It is a photochemical reaction.
- It takes place in the grana of chloroplasts.
- It requires participation of both PS I and PS II.
- The transfer of electrons through the ETS is unidirectional or non-cyclic.
- The non-cyclic process involves:
  - Photophosphorylation (ATP formation)
  - Photolysis of water
  - Formation of assimilatory power (ATP and $NADPH_2$)
  - Liberation of oxygen

The non-cyclic photophosphorylation takes place as follows:

Photoexcitation of PS I

The chlorophyll-a (P 700) of PS I is activated on receiving photons of light and so it expels electrons. As a result of this loss, it becomes ionized chl-a. The electrons from PS I are first accepted by FRS (ferredoxin reducing substance) and are transferred to co-enzyme NADP via Fd (ferredoxin). NADP retains the electrons and is thus reduced.

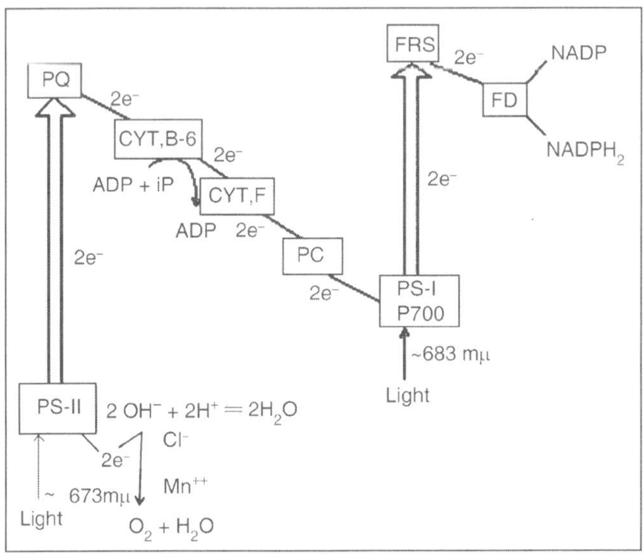

Fig. Diagrammatic Representation of the Non-cyclic Process (Z Scheme)

Photoexcitation of PS II

Similarly, on receiving photons of light, the chl-a (680) of PS II also gets activated and expels electrons. As a result, it becomes ionized chl-a. The high energy containing electrons from PS II are first accepted by an electron acceptor PQ (plastoquinone) and are then transferred along stet Electron Transport System consisting of plastoquinone, cytochrome b6, Cytochrome f and plastocyanin. Finally, the electrons are accepted by the ionized chl-a of PS I. In this way, ionized PS I comes back to the ground state and can again participate in the process.

Photolysis of Water

The ionized chlorophyll-a of PS II is brought back to the ground state with the help of the electrons made available through the photolysis of water.

$$2H_2O \xrightarrow[\text{Chlorophyll}]{\text{light}} 4e^- + 4H^+ + O_2 \left(\text{Photolysis of water}\right)$$

Formation of NADPH$_2$ and Liberation of O$_2$

During photolysis of water, oxygen is released as a by-product. The 2 NADP receive 4e from PS I and 4H from the photolysis of water to form reduced coenzyme NADPH$_2$

$$2NADP + 4e^- + 4H^+ \rightarrow 2NADPH_2$$

Non-Cyclic Electron Transfer

In this process, the movement of electrons is considered to be non-cyclic, or unidirectional. This is because, the final electron acceptor is different from the initial electron donor.

Initial electron donor Final electron acceptor:

(i) $Chl - a \text{ of } PSI \xrightarrow{e^-} NADP$

(ii) $Chl - a \text{ of } PSII \xrightarrow{e^-} Chl.a + \text{of } PSI$

(iii) $\underset{(Photolysis)}{Water} \xrightarrow{e^-} Chl.a + \text{ of } PSII$

In this way, the light-induced electron flow is maintained continuously in the non-cyclic manner from water to NADP as follows

$$Water \xrightarrow{e^-} PSII \xrightarrow{e^-} PSI \xrightarrow{e^-} NADP$$

Because of the zig-zag path of the electrons, the non cyclic process is also called 'Z' scheme electron transport.

ATP Formation

During the non-cyclic transfer of energy-rich electrons through the Electron Transport System from PS II to PS I, energy from the electrons is released. This is used for the formation of ATP from ADP and H_3PO_4. This is called non-cyclic photophosphorylation.

End Products of Non-cyclic Process and the Assimilatory Power

The end products are ATP, $NADPH_2$ and O_2. Of these, O_2 is liberated from the green plants. ATP and $NADPH_2$ are used in the following dark reactions (phase II) for carbon assimilation(*i.e.*, reduction of CO_2 to form carbohydrates). ATP and $NADPH_2$ of light reaction are called 'assimilatory power.

PHOTOSYNTHETIC ORGANISMS

The photosynthetic process in all plants and algae as well as in certain types of photosynthetic bacteria involves the reduction of CO_2 to carbohydrate and removal of electrons from H_20, which results in the release of O_2. In this process, known as oxygenic photosynthesis, water is oxidized by the photosystem II reaction centre, a multisubunit protein located in the photosynthetic membrane.

Years of research have shown that the structure and function of photosystem II is similar in plants, algae and certain bacteria, so that knowledge gained in one species can be applied to others. This homology is a common feature of proteins that perform the same reaction in different species. This homology at the molecular level is important because there are estimated to be 300,000-500,000 species of plants. If different species had evolved diverse mechanisms for oxidizing water, research aimed at a general understanding of photosynthetic water oxidation would be hopeless.

Anoxygenic Photosynthetic Organisms

Some photosynthetic bacteria can use light energy to extract electrons from molecules other than water. These organisms are of ancient origin, presumed to have evolved before oxygenic photosynthetic organisms. Anoxygenic photosynthetic organisms occur in the domain Bacteria and have representatives in four phyla - Purple Bacteria, Green Sulfur Bacteria, Green Gliding Bacteria, and Gram Positive Bacteria.

ROLE OF PHOTOSYNTHESIS

Photosynthesis is the physico-chemical process by which plants, algae and photosynthetic bacteria use light energy to drive the synthesis of organic compounds. In plants, algae and certain types of bacteria, the photosynthetic process results in the release of molecular oxygen and the removal of carbon dioxide from the atmosphere that is used to synthesize carbohydrates (oxygenic photosynthesis).

Other types of bacteria use light energy to create organic compounds but do not produce oxygen (anoxygenic photosynthesis). Photosynthesis provides the energy and reduced carbon required for the survival of virtually all life on our planet, as well as the molecular oxygen necessary for the survival of oxygen consuming organisms1.In addition, the fossil fuels currently being burned to provide energy for human activity were produced by ancient photosynthetic organisms. Although photosynthesis occurs in cells or organelles that are typically only a few microns across, the process has a profound impact on the earth's atmosphere and climate.

Each year more than 10% of the total atmospheric carbon dioxide is reduced to carbohydrate by photosynthetic organisms. Most, if not all, of the reduced carbon is returned to the atmosphere as carbon dioxide by microbial, plant and animal metabolism, and by biomass combustion. In turn, the performance of photosynthetic organisms depends on the earth's atmosphere and climate.

Over the next century, the large increase in the amount of atmospheric carbon dioxide created by human activity is certain to have a profound impact on the performance and competition of photosynthetic organisms. Knowledge of the physico-chemical process of photosynthesis is essential for understanding the relationship between living organisms and the atmosphere and the balance of life on earth. Several books on photosynthesis are available for the uninitiated or advanced student. Taiz and Zeiger place the photosynthetic process in the context of over all plant physiology, and Cramer and Knaff describe the bioenergetic foundation of photosynthesis.

The overall equation for photosynthesis is deceptively simple. In fact, a complex set of physical and chemical reactions must occur in a coordinated manner for the synthesis of carbohydrates. To produce a sugar molecule such as sucrose, plants require nearly 30 distinct proteins that work within a complicated membrane structure.

Research into the mechanism of photosynthesis centers on understanding the structure of the photosynthetic components and the molecular processes that use radiant energy to drive carbohydrate synthesis. The research involves several disciplines, including physics, biophysics, chemistry, structural biology, biochemistry, molecular biology and physiology, and serves as an outstanding example of the success of multidisciplinary research. As such, photosynthesis presents a special challenge in understanding several interrelated molecular processes.

BRIEF HISTORY

In the 1770s Joseph Priestley, an English chemist and clergyman, performed experiments showing that plants release a type of air that allows combustion. He demonstrated this by burning a candle in a closed vessel until the flame went out. He placed a sprig of mint in the chamber and after several days showed that the candle could burn again. Although Priestley did not know about molecular oxygen, his work showed that plants release oxygen into the atmosphere. It is noteworthy that over 200 years later, investigating the mechanism by which plants produce oxygen is one of the most active areas of photosynthetic research.

Building on the work of Priestley, Jan Ingenhousz, a Dutch physician, demonstrated that sunlight was necessary for photosynthesis and that only the green parts of plants could release oxygen. During this period Jean Senebier, a Swiss botanist and naturalist, discovered that CO_2 is required for photosynthetic growth and Nicolas- The odore de Saussure, a Swiss chemist and plant physiologist, showed that water is required. It was not until 1845 that Julius Robert von Mayer, a German physician and physicist, proposed that photosynthetic organisms convert light energy into chemical free energy. An interesting time line of the history of photosynthesis has been presented by Huzisige and Ke. By the middle of the nineteenth century the key features of plant photosynthesis were known, namely, that plants could use light energy to make carbohydrates from CO_2 and water. The empirical equation representing the net reaction of photosynthesis for oxygen evolving organisms is :

$$CO_2 + 2H_2O + \text{Light Energy} \rightarrow [CH_2O] + O_2 + H_2O, \quad (1)$$

where [CH2O] represents a carbohydrate (e.g., glucose, a six-carbon sugar). The synthesis of carbohydrate from carbon and water requires a large input of light energy. The standard free energy for the reduction of one mole of CO_2 to the level of glucose is +478 kJ/mol. Because glucose, a six carbon sugar, is often an intermediate product of photosynthesis, the net equation of photosynthesis is frequently written as:

$$6CO_2 + 12H_2O + \text{Light Energy} \rightarrow C_6H_{12}O_6 + 6O_2 + 6H_2O. \quad (2)$$

The standard free energy for the synthesis of glucose is +2,870 kJ/mol. Not surprisingly, early scientists studying photosynthesis concluded that the O_2 released by plants came from CO_2, which was thought to be split by light energy. In the 1930s comparison of bacterial and plant photosynthesis lead Cornelis van Niel to propose the general equation of photosynthesis that applies to plants, algae and photosynthetic bacteria. Van Niel was aware that some photosynthetic bacteria could use hydrogen sulfide (H_2S) instead of water for photosynthesis and that these organisms released sulfur instead of oxygen. Van Niel, among others, concluded that photosynthesis depends on electron donation and acceptor reactions and that the O_2 released during photosynthesis comes from the oxidation of water. Van Niel's generalized equation is :

$$CO_2 + 2H_2A + \text{Light Energy} \rightarrow [CH_2O] + 2A + H_2O. \quad (3)$$

In oxygenic photosynthesis, 2A is O_2, whereas in anoxygenic photosynthesis, which occurs in some photosynthetic bacteria, the electron donor can be an inorganic hydrogen donor, such as H2S (in which case A is elemental sulfur) or an organic hydrogen donor such as succinate (in which case, A is fumarate). Experimental evidence that molecular oxygen came from water was provided by Hill and Scarisbrick (1940) who demonstrated oxygen evolution in the absence of CO_2 in illuminated chloroplasts and by Ruben et al. (1941) who used 18O enriched water.

The biochemical conversion of CO_2 to carbohydrate is a reduction reaction that involves the rearrangement of covalent bonds between carbon, hydrogen and oxygen. The energy for the reduction of carbon is provided by energy rich molecules that are produced by the light driven electron transfer reactions. Carbon reduction can occur in the dark and involves a series of biochemical reactions that were elucidated by Melvin Calvin, Andrew Benson and James Bassham in the late 1940s and 1950s. Using the radioisotope 14C, most of the intermediate steps that result in the production of carbohydrate were identified. Calvin was awarded the Nobel Prize for Chemistry in 1961 for this work.

In 1954 Daniel Arnon and coworkers discovered that plants, and A. Frenkel discovered that photosynthetic bacteria, use light energy to produce ATP, an organic molecule that serves as an energy source for many biochemical reactions. During the same period L.N.M. Duysens showed that the primary photochemical reaction of photosynthesis is an oxidation/reduction reaction that occurs in a protein complex (the reaction centre). Over the next few years the work of several groups, including those of Robert Emerson, Bessel Kok, L.N.M. Duysens, Robert Hill and Horst Witt, combined to prove that plants, algae and cyanobacteria require two reaction centers, photosystem II and photosystem I, operating in series.

In 1961 Peter Mitchell suggested that cells can store energy by creating an electric field or a proton gradient across a membrane. Mitchell's proposal that energy is stored as an electrochemical gradient across a vesicular membrane opened the door for understanding energy transformation by membrane systems. He was awarded the Nobel Prize in Chemistry in 1978 for his theory of chemiosmotic energy transduction.

Most of the proteins required for the conversion of light energy and electron transfer reactions of photosynthesis are located in membranes. Despite decades of work, efforts to determine the structure of membrane bound proteins had little success. This changed in the 1980s when Johann Deisenhofer, Hartmut Michel, Robert Huber and co-workers determined the structure of the reaction centre of the purple bacterium Rhodospeudomonas viridis.. They were awarded the Nobel Prize for Chemistry in 1988 for their work, which has provided insight into the relationship between structure and function in membrane-bound proteins.

A key element in photosynthetic energy conversion is electron transfer within and between protein complexes and simple organic molecules. The electron transfer reactions are rapid (as fast as a few picoseconds) and highly specific. Much of our current understanding of the physical principles that guide electron transfer is based on the pioneering work of Rudolph A. Marcus, who received the Nobel Prize in Chemistry in 1992 for his contributions to the theory of electron transfer reaction in chemical systems.

PHOTOSYNTHESIS AND NUTRIENTS

We humans receive all the nourishment we need either directly from plants or else from herbivorous animals. Sunlight is simply a source of energy; but its raw form is not that practical. It is impossible to consume this energy, or to use it or store it directly in the body. Therefore, solar energy needs to be converted into another, usable form of energy—which is what photosynthesis does. By means of this process, plants turn solar energy into a stored form they can use at some later time. Using solar energy, the photosynthetic reaction centers in leaves converts the carbon dioxide in the air into starch and other high-energy carbohydrates.

The O_2 released after carbon dioxide has been used is released into the atmosphere. When the plant later requires energy, it uses the energy it has stored in these carbohydrates. Living things that feed on these plants, in turn, meet their own energy needs through the carbohydrates stored in them, by way of photosynthesis.

As we shall be seeing, photosynthesis is an exceedingly complex process. The fact that all living things acquire the food they need to live as a result of such a complex process is the work of the infinite knowledge and wisdom of Allah:

O humanity! Remember Allah's blessing to you. Is there any creator other than Allah providing for you from heaven and earth? There is no Allah but Him. So how have you been perverted? (Surah Fatir: 3)

PHOTOSYNTHESIS AND ENERGY

Your car engine runs on what was once solar energy. Jet planes fly thanks to what was once solar energy. You are using what was once solar energy even as you read these lines.

The first thing that comes into your mind as you read this will be that your car actually runs on gasoline and that jetliners use aviation fuel. You will imagine that you acquired the energy to read these lines from the last meal you ate, not from solar energy. The fact is, however, that both gas and the meals you eat, and even the wood and coal we burn for warmth, all contain energy obtained from the Sun—via photosynthesis. How?

Plants that stored solar energy in their bodies millions of years ago by means of photosynthesis, and animals that ate these plants, gave rise to the petroleum with which we are familiar, underground and under high pressure, after a period of millions of years. Coal and natural gas came into being in the same way. In short, the solar energy stored in plants thanks to photosynthesis was placed at our disposal after millions of years had passed.

Similarly, the energy you obtain from the food you eat is simply the solar energy stored by plants. The energy you obtain from animal foodstuffs is also the energy those animals obtained from plants. The source of energy is always the Sun, and the system that makes this energy useable by human beings is always photosynthesis. You cannot acquire the energy you possess by means of any system other than photosynthesis.

PHOTOSYNTHESIS AND SIDE PRODUCTS

Wood is a very important material used not just as fuel but in many fields, including construction. Paper, cotton and other natural fibers, for instance, consist of cellulose produced almost entirely by photosynthesizing plants. Even wool depends on the energy sheep obtain from grass, and photosynthesis. Solar energy, transformed by photosynthesis, is the source of countless vegetable, animal and organic side products.

PHOTOSYNTHESIS AND THE ENVIRONMENT

Living things constantly increase the carbon dioxide in the air and the air temperature. Millions of tons of carbon dioxide are released into the atmosphere every year as a result of the respiration of human beings, animals and micro-organisms in the soil, not to mention burning of fossil fuels. Moreover, the amount of carbon dioxide released into the atmosphere from fuels consumed for heating purposes in factories and homes and used in transport also reaches billions of tons. According to one study, an increase of 42 billion metric tons in the carbon dioxide has been seen in the atmosphere in the last 22 years. One of the main causes of this rise is the fuels consumed and deforestation. The rise in carbon dioxide caused by fuels over the last 22 years is 78 billion metric tons.

Unless this rise is compensated, there will be terrible ecological imbalances. In such an event, the amount of oxygen in the atmosphere will fall to very low levels, and the Earth's temperature will rise, as a result of which the ice caps will melt. Therefore, some low-lying regions will be flooded, while others will turn into deserts, and all life on Earth will be endangered. Yet that will not be the case, because carbon dioxide is constantly consumed and oxygen released in the process of photosynthesis carried out by plants and micro-organisms.

Thus the equilibrium in nature can persist. While there has been a rise of 78 billion tons in carbon dioxide caused by fuels, the level remaining in the atmosphere is 42 billion tons. This excess CO_2 is to a large extent cleaned up from the atmosphere by way of photosynthesis and the oceans.

The Earth's temperature is fixed within a specific range. Wide temperature fluctuations do not occur, because green plants also ensure temperature balance.

Photosynthesis ensures the continuity of these balances, which are of such vital importance for life on Earth. Also, there is no other mechanism for the preservation of the level of oxygen in the atmosphere.

PHOTOSYNTHESIS: LIGHT INTENSITY AND TEMPERATURE

As has been mentioned, the complex mechanism of photosynthesis includes a photochemical, or light-dependent, stage and an enzymatic, or dark, stage that involves chemical reactions. These stages can be distinguished by studying the rates of photosynthesis at various degrees of light saturation (*i.e.*, intensity) and at different temperatures. Over a range of moderate temperatures and at low to medium light intensities (relative to the normal range of the plant species), the rate of photosynthesis increases as the intensity increases and is independent of temperature.

As the light intensity increases to higher levels, however, the rate becomes increasingly dependent on temperature and less dependent on intensity; light "saturation" is achieved at a specific light intensity, and the rate then is dependent only on temperature if all other factors are constant. In the light-dependent range before saturation, therefore, the rate of photosynthesis is determined by the rates of photochemical steps. At high light intensities, some of the chemical reactions of the dark stage become rate-limiting. At light saturation, rate increases with temperature until a point is reached beyond which no further rate increase can occur.

In many land plants, moreover, a process called photorespiration occurs at high light intensities and temperatures. Photorespiration competes with photosynthesis and limits further increases in the rate of photosynthesis, especially if the supply of water is limited.

Photosynthesis: Carbon Dioxide

Included among the rate-limiting steps of the dark stage of photosynthesis are the chemical reactions by which organic compounds are formed using carbon dioxide as a carbon source. The rates of these reactions can be increased somewhat by increasing the carbon dioxide concentration. During the past century, the level of carbon dioxide in the atmosphere has been rising due to the extensive combustion of fossil fuels.

The atmospheric level of carbon dioxide climbed from about 0.028 per cent in 1860 to 0.0315 per cent by 1958 (when improved measurements began), and to 0.034 per cent by 1981. This increase in carbon dioxide directly increases plant photosynthesis, but the size of the increase depends on the species and physiological condition of the plant. Furthermore, if increasing levels of atmospheric carbon dioxide result in climatic changes, including increased global temperatures as some meteorologists predict, these changes will affect photosynthesis rates.

Factors That Influence the Rate of Photosynthesis:Water

For land plants, water availability can function as a limiting factor in photosynthesis and plant growth. Besides the requirement for water in the photosynthetic reaction itself, water is transpired from the leaves; that is, water evaporates from the leaves to the atmosphere via the stomates. These stomates are small openings through the leaf epidermis, or outer skin; they permit the entry of carbon dioxide but also allow the exit of water vapour. The stomates open and close according to the physiological needs of the leaf. In hot and arid climates the stomates may close to conserve water, but this closure limits the entry of carbon dioxide and hence the rate of photosynthesis, while the wasteful process of photorespiration may increase. If the level of carbon dioxide in the atmosphere increases, more carbon dioxide could enter through a smaller opening of the stomates, so that more photosynthesis could occur with a given supply of water.

Factors that Influence the Rate of Photosynthesis: Minerals

Several minerals are required for healthy plant growth and for maximum rates of photosynthesis. Nitrate or ammonia, sulfate, phosphate, iron, magnesium, and potassium are required in substantial amounts for the synthesis of amino acids, proteins, coenzymes, deoxyribonucleic acid (DNA) and ribonucleic acid (RNA), chlorophyll and other pigments, and other essential plant constituents.

Smaller amounts of such elements as manganese, copper, and chlorine are required in photosynthesis. Some other trace elements are needed for various nonphotosynthetic functions in plants.

Factors that Influence the Rate of Photosynthesis: Internal Factors

Each plant species adapts to a range of environmental factors. Within this normal range of conditions, complex regulatory mechanisms in the plant's cells adjust the activities of enzymes (*i.e.,* organic catalysts). These adjustments maintain a balance in the overall photosynthetic process and control it in accordance with the needs of the whole plant. With a given plant species, for example, doubling the carbon dioxide level might cause a temporary increase of nearly twofold in the rate of photosynthesis; a few hours later, however, the rate might fall to the original level because photosynthesis had made more sucrose than the rest of the plant could use. By contrast, another plant species provided with such carbon dioxide enrichment might be able to use more sucrose and would continue to photosynthesize and to grow faster throughout most of its life cycle.

THE CONVERSION OF LIGHT ENERGY TO ATP

The electron transfers of the light reactions provide the energy for the synthesis of two compounds vital to the dark reactions: NADPH> and ATP. The previous section explained how noncyclic electron flow results in the reduction of NADP to NADPH. In this section, the synthesis of the energy-rich compound ATP is described.

ATP is formed by the addition of a phosphate group to a molecule of adenosine diphosphate (ADP); or to state it in chemical terms, by the phosphorylation of ADP. This reaction requires a substantial input of energy, much of which is captured in the bond that links the added phosphate group to ADP. Because light energy powers this reaction in the chloroplasts, the production of ATP during photosynthesis is referred to as photophosphorylation.

Unlike the production of NADPH, the photophosphorylation of ADP occurs in conjunction with both cyclic and noncyclic electron flow. In fact researchers speculate that the sole purpose of cyclic electron flow may be for photophosphorylation, since this process involves no net transfer of electrons to reducing agents. The relative amounts of cyclic and noncyclic flow may be adjusted in accordance with changing physiological needs for ATP and reduced ferrodoxin and NADPH in chloroplasts. In contrast to electron transfer in light reactions I and II, which can occur in membrane fragments, intact thylakoids are required for efficient photophosphorylation. This requirement stems from the special nature of the mechanism linking photophosphorylation to electron flow in the lamellae.

The theory relating the formation of ATP to electron flow in the membranes of both chloroplasts and mitochondria (the organelles responsible for ATP formation during cellular respiration) was first proposed by the English biochemist Peter Mitchell. This chemiosmotic theory has been somewhat modified to fit later experimental facts, and there is still debate over many of the details. The general features, however, are widely accepted. A central feature is the formation of a hydrogen ion (proton) concentration gradient and an electrical charge across intact lamellae. The potential energy stored by the proton gradient and electrical charge is then used to drive the energetically unfavourable conversion of ADP and inorganic phosphate (P_i) to ATP and water.

The manganese-protein complex associated with light reaction II is exposed to the interior of the thylakoid. Consequently, the oxidation of water during light reaction II leads to release of hydrogen ions (protons) into the inner thylakoid space. Furthermore, it is likely that photoreaction II entails the transfer of electrons across the lamella toward its outer face, so that when plastoquinone molecules are reduced they can receive protons from the outside of the thylakoid.

When these reduced plastoquinone molecules are oxidized, giving up electrons to the cytochrome-iron-sulfur complex, protons are released inside the thylakoid. Because the lamella is impermeable to them, the release of protons inside the thylakoid by oxidation of both water and plastoquinone leads to a higher concentration of protons inside the thylakoid than outside it.

In other words, a proton gradient is established across the lamella. The movement of electrons (negatively charged particles) outward across the lamella during both light reactions results in the establishment of an electrical charge across the lamella. (Some scientists believe, however, that the proton gradient and electrical charge required for ATP formation need not be between inner and outer thylakoid space but only within the membrane.)

An enzyme complex located partly in and on the lamellae catalyzes the reaction in which ATP is formed from ADP and inorganic phosphate. The reverse of this reaction is catalyzed by an enzyme called ATP-ase, hence the enzyme complex is sometimes called an ATP-ase complex. It is also called the coupling factor. It consists of hydrophilic polypeptides (F_1), which project from the outer surface of the lamellae, and hydrophobic polypeptides (F_0), which are embedded inside the lamellae. Researchers hypothesize that F_0 forms a channel that permits protons to flow through the lamellar membrane to F_1. The enzymes in F_1 then catalyze ATP formation, using both the proton supply and the lamellar transmembrane charge.

In summary, the use of light energy for ATP formation occurs indirectly: a proton gradient and electrical charge—built up in or across the lamellae as a consequence of electron flow in the light reactions—provide the energy to drive the synthesis of ATP from ADP and P_i.

CARBON FIXATION AND REDUCTION

The assimilation of carbon into organic compounds is the result of a complex series of enzymatically regulated chemical reactions—the dark reactions. This term is something of a misnomer, for these reactions can take place in either light or darkness. Furthermore, some of the enzymes involved in the so-called dark reactions become inactive in prolonged darkness.

Elucidation of the Carbon Pathway

Radioactive isotopes of carbon (C) and phosphorus (P) have been valuable in identifying the intermediate compounds formed during carbon assimilation. A photosynthesizing plant does not strongly discriminate between the natural carbon isotopes and C. During photosynthesis in the presence of CO_2, the compounds formed become labelled with the radioisotope. During very short exposures, only the first intermediates in the carbon-fixing pathway become labelled. Early investigations showed that some radioactive products were formed even when the light was turned off and the CO_2 was added just afterward in the dark, confirming the nature of the carbon fixation as a "dark" reaction.

The U.S. biochemist Melvin Calvin, a Nobel Prize recipient for his work on the carbon reduction cycle, allowed green plants to photosynthesize in the presence of radioactive carbon dioxide for a few seconds under various experimental conditions. Products that became labelled with radioactive carbon during Calvin's experiments included a three-carbon compound called 3-phosphoglycerate, sugar phosphates, amino acids, sucrose, and carboxylic acids. When photosynthesis was stopped after two seconds, the principal radioactive product was PGA, which therefore was identified as the first compound formed during carbon dioxide fixation in green plants.

Further studies with C as well as with inorganic phosphate labelled with P led to the mapping of the carbon fixation and reduction pathway called the reductive pentose phosphate cycle (RPP cycle). An additional pathway for carbon transport in certain plants was later discovered in other laboratories. All the steps in these pathways can be carried out in the laboratory by isolated enzymes in the dark. Several steps require the ATP or NADPH generated by the light reactions. In addition, some of the enzymes are fully active only when conditions simulate those in green cells exposed to light. In vivo, these enzymes are active during photosynthesis but not in the dark.

The Reductive Pentose Phosphate Cycle

Overall reaction. The RPP cycle, in which carbon is fixed, reduced, and utilized, involves the formation of intermediate sugar phosphates in a cyclic sequence (). One complete RPP cycle incorporates three molecules of carbon dioxide and produces one molecule of the three-carbon compound glyceraldehyde-3-phosphate (Gal3P). This three-carbon sugar phosphate usually is either exported from the chloroplasts or is converted to starch (dashed lines).

Summarizes the main steps that take place during one complete turn of the RPP cycle. ATP and NADPH formed during the light reactions are utilized for key steps in this pathway and provide the energy and reducing equivalents (*i.e.*, electrons) to drive the sequence in the direction shown. For each molecule of carbon dioxide that is fixed, two molecules of NADPH and three molecules of ATP from the light reactions are required. The overall reaction can be represented as follows:

$$9ATP + 6NADPH + 3CO_2 \rightarrow Gal3P + 6NADP^+ + 9ADP + 8P_i$$

The cycle is composed of four stages:

- Carboxylation,
- Reduction,
- Isomerization/condensation/dismutation, and
- Phosphorylation.

Carboxylation

The initial incorporation of carbon dioxide, which is catalyzed by the enzyme ribulose 1,5-bisphosphate carboxylase, proceeds by the addition of carbon dioxide to the five-carbon compound ribulose 1,5-bisphosphate (RuBP) and the splitting of the resulting unstable six-carbon compound into two molecules of PGA, a three-carbon compound. As indicated in, this reaction occurs three times during each complete turn of the cycle; thus, six molecules of PGA are produced.

Reduction

The six molecules of PGA are first phosphorylated with ATP by the enzyme PGA-kinase, yielding six molecules of 1,3-diphosphoglycerate (DPGA). These then are reduced with NADPH and the enzyme glyceraldehyde-3-phosphate dehydrogenase to give six molecules of Gal3P. These reactions are the reverse of two steps of the process glycolysis in cellular respiration.

Isomerization/condensation/dismutation

For each complete RPP cycle, one of the Gal3P molecules, with its three carbon atoms, is the net product and may be transferred out of the chloroplast or converted to starch inside the chloroplast (dashed lines,). For the cycle to regenerate, the other five Gal3P molecules (with a total of 15 carbon atoms) must be converted back to three molecules of five-carbon RuBP. The conversion of Gal3P to RuBP begins with a complex series of enzymatically regulated reactions (not shown in) that lead to the synthesis of the five-carbon compound ribulose-5-phosphate (Ru5P). The occurrence of this complex series, which entails isomerization, condensation, and dismutation reactions, is indicated by the broken arrow in.

Phosphorylation

The three molecules of Ru5P are converted to the carboxylation substrate,

RuBP, by the enzyme phosphoribulokinase, using ATP. This reaction, shown below, completes the cycle.

$$3Ru5P + 3ATP \rightarrow \; 3RuBP + 3ADP$$

REGULATION OF THE CYCLE

Photosynthesis cannot occur at night, but the respiratory process of glycolysis—which uses some of the same reactions as the RPP cycle, except in the reverse—does take place. Thus, some steps in the RPP cycle would be wasteful if allowed to occur in the dark because they would counteract the reactions of glycolysis. For this reason, some enzymes of the RPP cycle are "turned off " (*i.e.*, become inactive) in the dark.

Even in the presence of light, changes in physiological conditions frequently necessitate adjustments in the relative rates of reactions of the RPP cycle, so that enzymes for some reactions change in their catalytic activity. These alterations in enzyme activity typically are brought about by changes in levels of such chloroplast components as reduced ferredoxin, acids, and soluble components (*e.g.*, P_i and magnesium ions).

Products of carbon reduction

The most important use of Gal3P is its export from the chloroplasts to the cytoplasm of green cells, where it is used for biosynthesis of products needed by the plant. In land plants, a principal product is sucrose, which is translocated from the green cells of the leaves to other parts of the plant. Other key products include the carbon skeletons of certain primary amino acids, such as alanine, glutamate, and aspartate.

To complete the synthesis of these compounds, amino groups are added to the appropriate carbon skeletons made from Gal3P. Sulfur amino acids such as cysteine are formed by adding sulfhydryl groups and amino groups. Other biosynthesis pathways lead from Gal3P to lipids, pigments, and most of the constituents of green cells.

Starch synthesis and accumulation in the chloroplasts occurs particularly when photosynthetic carbon fixation exceeds the needs of the plant. Under such circumstances, sugar phosphates accumulate in the cytoplasm, binding cytoplasmic P_i. The export of Gal3P from the chloroplasts is tied to a one-for-one exchange of P_i for Gal3P, so less cytoplasmic P_i results in decreased export of Gal3P and decreased P_i in the chloroplast. These changes trigger alterations in the activities of regulated enzymes, leading in turn to increased starch synthesis. This starch can be broken down at night and used as a source of reduced carbon and energy for the physiological needs of the plant. Too much starch in the chloroplasts leads to diminished rates of photosynthesis, however. Thus, under what would seem to be the ideal photosynthetic conditions of a bright, warm day, many plants in fact have slower rates of photosynthesis in the afternoon.

Photorespiration

Under conditions of high light intensity, hot weather, and water limitation, the productivity of the RPP cycle is limited in many plants by the occurrence of photorespiration. This process converts sugar phosphates back to carbon dioxide; it is initiated by the oxygenation of RuBP (*i.e.,* the combination of gaseous oxygen [O_2] with RuBP). This oxygenation reaction yields only one molecule of PGA and one molecule of a two-carbon acid, phosphoglycollate, which is subsequently converted in part to carbon dioxide. The reaction of oxygen with RuBP is in direct competition with the carboxylation reaction (CO_2 + RuBP) that initiates the RPP cycle and is, in fact, catalyzed by the same protein, ribulose 1,5-bisphosphate carboxylase. The relative concentrations of oxygen and carbon dioxide within the chloroplasts determine whether oxygenation or carboxylation is favoured. The concentration of oxygen inside the chloroplasts may be higher than atmospheric (20 per cent) owing to photosynthetic oxygen evolution, whereas the internal carbon dioxide concentration may be lower than atmospheric (0.035 per cent) owing to photosynthetic uptake. Any increase in the internal carbon dioxide pressure tends to help the carboxylation reaction compete more effectively with oxygenation.

Carbon fixation via C_4 acids

Certain plants—including the important crops sugarcane and corn (maize), as well as other diverse species believed to have evolved in the drier tropical areas—have developed a special mechanism of carbon fixation that largely prevents photorespiration. The leaves of these plants have special anatomy and biochemistry. In particular, photosynthetic functions are divided between mesophyll and bundle sheath leaf cells. The carbon fixation pathway begins in the mesophyll cells, where carbon dioxide is added to the three-carbon acid phosphoenolpyruvate (PEPA) by an enzyme called phosphoenolpyruvate carboxylase. The product of this reaction is the four-carbon acid oxaloacetate, which is reduced to malate, another four-carbon acid, in one form of C_4 pathway. Malate then is translocated to bundle sheath cells, which are located near the vascular system of the leaf. There, malate enters the chloroplasts and is oxidized and decarboxylated (*i.e.,* loses CO_2) by malic enzyme. This yields carbon dioxide, which is fed into the RPP cycle of the bundle sheath cells, and pyruvate, a three-carbon acid that is translocated back to the mesophyll cells. In the mesophyll chloroplasts, the enzyme pyruvate orthophosphate dikinase (PPDK) uses ATP and P_i to convert pyruvate back to PEPA, completing the C_4 cycle. There are several variations of this pathway in different species. For example, the amino acids aspartate and alanine can substitute for malate and pyruvate in some species.

The C_4 pathway acts as a shuttle for carrying carbon dioxide into the chloroplasts of the bundle sheath cells, where it is used in carbohydrate synthesis. The resulting higher level of internal carbon dioxide in these chloroplasts serves to increase the ratio of carboxylation to oxygenation, thus minimizing photorespiration. Although the plant must expend extra energy to drive this shuttle, the energy loss is more than compensated by the near elimination of photorespiration under conditions where it would otherwise occur. Sugarcane and certain other plants that employ this pathway have the highest annual yields of biomass of all species.

THE MOLECULAR BIOLOGY OF PHOTOSYNTHESIS

Oxygenic photosynthesis occurs in both prokaryotic cells (cyanophytes) and eukaryotic cells (algae and higher plants). In eukaryotic cells, which contain chloroplasts and a nucleus, the genetic information needed for the reproduction of the photosynthetic apparatus is contained partly in the chloroplast chromosome and partly in chromosomes of the nucleus. For example, the carboxylation enzyme ribulose 1,5-bisphosphate carboxylase is a large protein molecule comprising a complex of eight large polypeptide subunits and eight small polypeptide subunits. The gene for the large subunits is located in the chloroplast chromosome, while the gene for the small subunits is in the nucleus. Transcription of the DNA of the nuclear gene yields messenger RNA (mRNA) that encodes the information for the synthesis of the small polypeptides. During this synthesis, which occurs on the cytoplasmic ribosomes, some extra amino acid residues are added to form a recognition leader on the end of the polypeptide chain. This leader is recognized by special receptor sites on the outer chloroplast membrane; these receptor sites then allow the polypeptide to penetrate the membrane and enter the chloroplast. The leader is removed and the small subunits combine with the large subunits, which have been synthesized on chloroplast ribosomes according to mRNA transcribed from the chloroplast DNA. The expression of nuclear genes that code for proteins needed in the chloroplasts appears to be under control of events in the chloroplasts in some cases; for example, the synthesis of some nuclear-encoded chloroplast enzymes may occur only when light is absorbed by chloroplasts.

DIVERSITY IN PHOTOSYNTHETIC PATHWAY

Various experiments and investigations regarding the fixation of CO_2 and the path of carbon during the dark reaction have indicated that there are a number of different pathways for CO_2 fixation in green plants.

These are:

- The C_3 pathway or Calvin cycle is the main pathway 4-C dicarboxylic acid called oxaloacetic acid (OAA). Such plants are called C_4 plants and the path of carbon (dark reaction) is the C_4 pathway. It was first noticed by Kortschak (1964) in the photosynthesis of sugarcane leaves. However, details of the C_4 pathway (*i.e.* the first CO_2 acceptor, the first stable product, the complete carbon pathway, etc.) were worked out by Hatch and Slack (1966). Therefore, it is known as Hatch-Slack Pathway. C_4 pathway is observed in many plants of family Gramineae (*e.g.* sugarcane, maize, some other monocot and some dicot plants).

- *Anatomical peculiarities of C_4 plants:* Most of the C_4 plants have a characteristic leaf anatomy and dimorphic chloroplasts. For example,
  - The leaf mesophyll consists of more or less compactly arranged cells.
  - It is not differentiated into palisade and spongy mesophyll as it is in C3 plants.
  - The vascular bundles (veins) in the leaf are surrounded by a distinct bundle sheath of radial enlarged parenchyma cells.
  - The chloroplasts in leaf cells are dimorphic(*i.e.*, of two types):
    (a) Chloroplasts in the mesophyll cells are smaller and possess grana.
    (b) Chloroplasts in the bundle sheath cells are larger and without grana.

This type of leaf anatomy in C_4 plants is described as Kranz anatomy.

Important Steps in Hatch and Slack Pathway

The various steps in C_4 pathway are completed in two parts and in two different regions in the leaves.

- First part reactions are completed in the stroma of the chloroplasts in mesophyll cells,
- Second part reactions are completed in the stroma of chloroplasts in bundle sheath cells.
- Part - I (in mesophyll cells)
  - *First CO_2 Fixation*: In the pathway, the first CO_2 acceptor is the 3-C phosphoenol pyruvate acid (PEP). CO_2 first combines with 3-C PEP to form 4-C OAA (oxaloacetic acid). As OAA is a dicarboxylic acid, this is also known as the dicarboxylic acid pathway.
  - 4-C OAA may be converted into 4-C malic acid or 4-C aspartic acid and transported to bundle sheath cells.
- Part - II (in bundle sheath cells)
  - In the chloroplasts of the bundle sheath cells, 4-C malic acid undergoes decarboxylation to form CO_2 and 3-C pyruvic acid.
  - *Second CO_2 fixation*: The CO_2 released in decarboxylation of malic acid combines with 5-C RUDP (ribulose diphosphate) to form 2 molecules of 3-C PGA as in the Calvin cycle. Further conversion of PGA to sugars is the same as in the Calvin cycle.
  - The pyruvic acid produced in decarboxylation of malic acid is transported back to the mesophyll cells. Here it is converted into PEPC and again made available for the C_4 pathway.

These steps in the Hatch-Slack pathway in the mesophyll and the bundle sheath cells are schematically shown in the figure below. Thus, in the C_4 plants, the initial few steps are different (typical of C_4 pathway). However, later on the reactions are similar to the Calvin cycle (*i.e.*, C_3 pathway). Hence, C_4 plants have both C_4 and C_3 pathways. Moreover, in C_4 plants, CO_2 fixation (*i.e.* carboxylation) occurs twice: first in the mesophyll cells (PEP + CO_2) and then again in the bundle sheath cells (RUDP + CO_2). For this reason, the C_4 pathway is also called a dicarboxylation pathway.

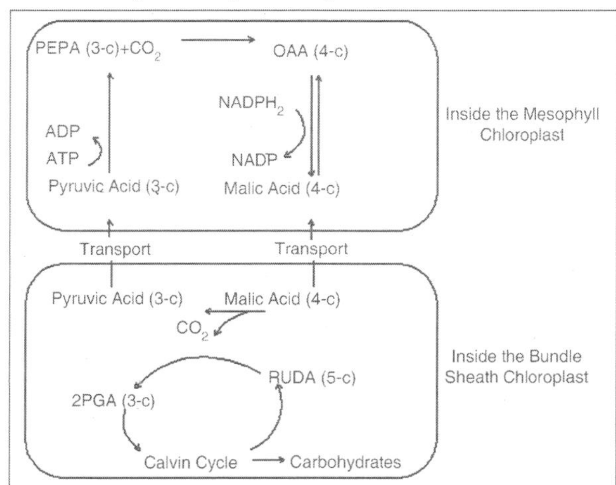

Fig. Schematic Representation of Hatch and Slackpathway

CAM Plants (Crassulacean Acid Metabolism)

The CAM plants are mostly succulent xerophytes such as plants of families Crassulaceae, Euphorbiaceae, Cactaceae, etc. The stomata in these plants remain closed during the day. This helps to check transpiration and open them during the cool night for gaseous exchange.

These plants are characterized by the presence of Crassulacean acid metabolism (CAM) involving night photosynthetic activity. This is a desert adaptation where stomata must remain closed during the day.

The CAM plants have the dark reaction very much similar to the C_4 plants (*i.e.*, they have both C_4 and C_3 pathways.) However, they differ from C_4 plants because, in CAM plants, both pathways occur in mesophyll cells only. Moreover, C_4 pathway occurs during the night when stomata are open (night photosynthetic activity) and C_3 pathway is completed during day time when the stomata are closed.

At night, atmospheric carbon dioxide is taken in through the open stomata. As in the C_4 pathway, carbon dioxide combines with 3-C PEP to form 4-C OAA. This is then converted to malic acid and stored in the cells. Thus, organic acids accumulate in the dark. During the day, when stomata are closed and atmospheric carbon dioxide is not available, the 4-C malic acid undergoes decarboxylation to form carbon dioxide and 3-C pyruvic acid. Carbon dioxide thus released enters the Calvin cycle (C_3 cycle) to form sugars.

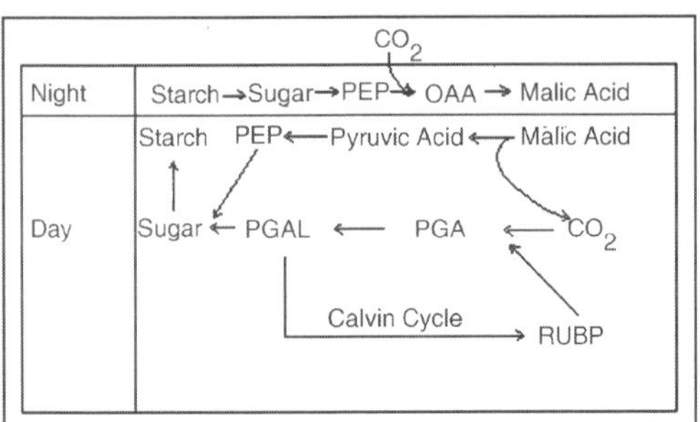

Fig. Schematic Representation of the CAM Pathway

Significance of Photosynthesis:

- It is the primary source of organic food and food energy (ATP) for all forms of life, either directly or indirectly.
- Excess sugars produced in photosynthesis are either stored in the form of carbohydrates or used in the biosynthesis of other organic compounds.
- In any ecosystem, green plants represent the most essential biotic components as they are the primary producers.
- Photosynthesis helps to purify air and also maintain balance of oxygen and carbon dioxide in the ecosystem.
- Oxygenic photosynthesis was responsible for converting the totally anaerobic condition on earth into aerobic atmosphere present now.
- The fossil fuels (*e.g.* natural gas, coal, petroleum (oil), etc.) are all energy-rich materials of an organic origin. The energy stored in all these fuels is basically solar energy which was trapped and stored during photosynthesis in the geological past.

REDOX SIGNALLING PATHWAYS

Present knowledge suggests that redox components initiate important signalling pathways, which ultimately regulate PS gene expression. Two questions are crucial for an understanding of this phenomenon at the molecular level.

- How are the redox signals transduced to their respective target genes?
- Do the various redox effects observed in different in vivo and in vitro systems reflect an integrated redox signaling network or do they represent separate signalling pathways, which operate independently from each other?

Perceptional redox signal transduction in oxygen-controlled. PS gene expression of purple bacteria is quite well understood, even if details remain to be clarified.

In contrast, our understanding of the signal systems under perceptional redox control starting from the PQ pool and:or the cyt *b*6 *f* complex is poor.

The large number of different organisms as well as physiological test systems makes it difficult to draw general conclusions, in particular, since species-specific mechanisms cannot be excluded at present. Nevertheless, common features and questions are obvious. First, how is the redox signal transferred out of the thylakoid membrane?

It was proposed that a putative factor, whose activity depends directly on the redox state of the QO site in cyt *b*6, activates PsaA:B synthesis. This seems to be supported by observations that show that the primary redox signal of the PQ pool is mediated via the cyt *b*6 *f* complex, but the latter investigations are focused on the redox activation of the LHCII kinase and not on redox effects on gene expression. A further potential candidate for the transduction of the PQ:cyt *b*6 *f* redox signal may be the thylakoid-associated kinase. It is still unclear whether state transitions and changes in chloroplast transcription are controlled by two different pathways or by two different branches initiated by the same redox sensor. However, in mustard, these two processes are responses to changes in the PQ redox state, which occur on the same time scale, i.e. both responses can be observed within minutes. This suggests that they act functionally in parallel with the state transition as the gate triggering the gene expression response.

It was also hypothesised that the redox state of the PQ is sensed by a membrane-bound, two-component sensor kinase, which transfers the signal to a response regulator, which, in turn, affects gene expression (Allen 1993a,b). In cyanobacteria and purple bacteria, two-component systems are known to be involved in PS gene expression (Allen 1993c). Recent evidence strongly implicates a redox twocomponent system in both state transitions and photosystem stoichiometry adjustment in the cyanobacterium *Synechocystis* 6803 (Li and Sherman 2000). Histidine sensor kinases have also been identified in the cytoplasm of higher plants, such as ethylene- and cytokinine-sensing kinases, and a still growing number of nuclear-encoded response-regulator-like genes have been identified in *Arabidopsis* and maize.

However, none of the identified genes (Imamura et al. 1999) show typical chloroplast transit peptides with the ChloroP programme. The functional and molecular analysis of plant response-regulator-like proteins is just at its beginning, and further studies will clarify their putative role in cellular redox signalling mechanisms.

In contrast, transductional control of the chloroplast gene *psbA* via the ferredoxin-thioredoxin system and the translation initiation complex in *Chlamydomonas* or via glutathione and the RNA polymerase kinase complex in mustard appears to be better understood.

Components for the transduction of the redox signal to the gene expression level are known and appropriate models of their action are proposed. Further studies will show if these models apply in vivo and if other plastid genes besides *psbA* are regulated in the same way. Redox control of the expression of nuclear PS genes represents a second major task in plant research, as the signal has to pass through the chloroplast envelope. The mechanisms by which the signal is transported over the membrane are completely unknown to date. In addition, it is not clear if there exist different redox signals and, if yes, how many. The various physiological test systems (e.g. light quantity or light quality), which can be used to investigate redox signals, suggest the existence of at least several signals.

In any case, redox signals may represent one form of the so-called 'plastid factor', whose nature is still unknown. As already mentioned, studies with *Dunaliella tertiolecta* point to the involvement of a phosphorylatable protein messenger in this interorganellar communication. Recent studies with the *cue*1 (*cab underexpressed*) mutant from *Arabidopsis* suggest also the involvement of a plastid phosphoenolpyruvate: phosphate translocator in this pathway.

As chloroplast metabolism and photosynthesis are tightly coupled, it is possible that both contribute to the chloroplast signal. An important difference in redox control of nuclear and chloroplast PS genes, at least in higher plants, is that the redox signal passing to the nucleus is composed of about 80–120 individual signals (depending on the total number of chloroplasts in the respective cell), while the redox control within each chloroplast is organelle-specific. As the redox poise of the chloroplasts within one cell may differ because of their cellular position, nuclear gene expression reflects the response to an average of all these signals.

Therefore, chloroplast redox signals to the nucleus contain more general information about the photosynthetic capacity of the cell, while redox signals within the plastid allow control of gene expression according to the specific situation within a single chloroplast. Answers to these questions could be obtained by comparison of redox pathways in cells with multiple chloroplasts (e.g. higher plants) and in cells with a single chloroplast, such as *Chlamydomonas* or the *ftsZ* mutant of *Arabidopsis*.

As outlined above, redox control in chloroplasts starts from different redox parameters. How do these various redox signals contribute to the gene-specific control of PS gene expression in a coordinated way? We still need more information to understand this mechanism. Nevertheless, it appears that a hierarchical order of the various redox.

The triangle represents an increase in incident light intensity from the left to the right. The light intensity and the respective main active redox control parameter are correlated by perpendicular lines representing 3 light intensity windows. Each parameter has to fulfil a different physiological role according to the light intensity, which in the end helps to acclimate photosynthesis to changing illumination. Signals influences gene expression according to the incident light intensity.

Our general assumption is that a redox control parameter is active only under those light intensities under which it shows variable ratios in its oxidized versus reduced forms. A model is based on 3 light intensity windows, i.e. low, moderate and high light intensity. The range of the low light intensity is defined by the redox state of the PQ pool, which depends directly on the linear electron transport. Its main physiological role is to redistribute imbalances in energy excitation between the two photosystems to provide an efficient electron flux, even under limited photon yields, such as under water, in shade or under cloud, by means of activation of physiological variations or gene expression mechanisms.

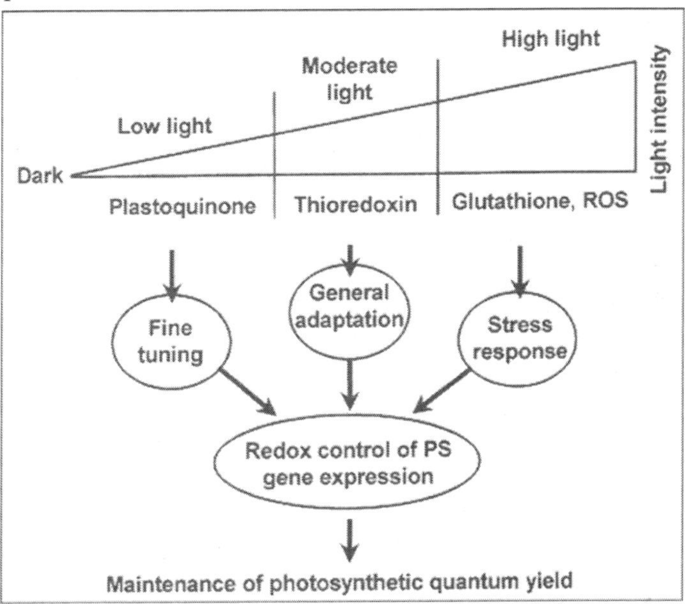

Fig. Simplified Model of Hierarchical Order in Light-Dependent Redox control of PS gene expression in a photosynthetic cell.

Reaching the point of complete PQ reduction by increasing light intensity, the second window becomes activated, in which we suggest that thioredoxin may become the predominant redox control parameter. Thioredoxin is reduced by ferredoxin via a ferredoxin-thioredoxin reductase, a crucial step in photosynthesis redox chemistry, since the received electron signal from ferredoxin is transduced into a thiol signal transmitted to thioredoxin. The redox state of thioredoxin may still be variable under moderate light intensities, as it depends on the linear and cyclic electron transport. In addition, the midpoint redox potential (Em) of thioredoxin (290 mV for thioredoxin f and –300 mV for thioredoxin m, Schu¨ rmann and Jacquot 2000) is more negative than that of PQ (100 mV). According to the Nernst equation, under a constant pH and equal total concentration (of PQ and thioredoxin), PQ has a ratio of reduced to oxidized forms of 107 when at equilibrium with thioredoxin that is exactly 50% reduced.

Differences in the redox potential of the enzymes of the Calvin cycle suggest a sequential activa- tion of these enzymes by the reduction state of thioredoxin as well as different degrees of activation. Therefore, thioredoxin may act in a similar way as regulator of gene expression. The third proposed window describes conditions in which the incident light intensity increases such that thioredoxin remains reduced. As a consequence, dissipation of excess light energy becomes important in order to avoid photooxidative damage. High light stress therefore increases the GSSG:GSH ratio as well as ROS production, such as H_2O_2. Increasing concentrations of GSSG or H_2O_2 then act as the activation signals for light stress defence mechanisms.

The width of each light intensity window depends on the photosynthetic capacity of the individual organism, which explains why a given gene under the same conditions is controlled by different redox mechanisms in different species. It is also important to note that the action of the 3 proposed redox mechanisms may overlap to a certain degree.

TRANSDUCTIONAL REDOX CONTROL

Transductional redox control occurs if the perceived stimulus leads to a change in the redox state of a downstream molecule of the sensory system. This type of redox control has been described for unicellular algae and higher plants. Studies in many laboratories focus on redox agents, which are reduced or oxidised by the electron transport chain (such as thioredoxin or glutathione) and which ultimately influence components involved in PS gene expression. Experimentally, different redox agents are applied to activity tests with the purified enzyme of interest. While these data are based on in vitro observations, their physiological importance often remains to be determined.

In the unicellular green alga *Chlamydomonas reinhardtii*, it was shown that binding of a translation-activating protein complex to the *psbA* mRNA 5%-untranslated region is enhanced under reducing conditions in vitro, probably by means of activation by reduced thioredoxin. Further studies suggested the involvement of the disulphide isomerase-like enzyme RB60 as transducing redox active enzyme activity. The first physiological support for this redox pathway came from a PSI-deficient mutant, which shows a reduced *psbA* translation. Recent in organello experiments show that light is indeed able to reduce RB60 and to enhance *psbA* translation.

These results support the idea of light-controlled *psbA* translation via the ferredoxin-thioredoxin system. In addition, electron transport efficiency also appears to influence the splicing of the *psbA* message in *Chlamydomonas*, but the present data do not distinguish between perceptional and transductional redox mechanisms. Finally, the application of DCMU or redox agents to *Chlamydomonas* chloroplast transformant lines MU7 grown under dark:light cycles indicates a regulatory role of photosynthetic electron transport and redox state on the degradation of chloroplast transcripts, probably via the ferredoxin-thioredoxin system.

In higher plants, it was shown that the activity of the plastid RNA processing enzyme p54 from mustard is influenced by reducing agents in vitro.

The same is true for the plastid transcription kinase from mustard, which catalyses phosphorylation of sigmalike factors in vitro. PTK activity is enhanced in vitro by reduced glutathione. Glutathione is thought to play an important role in enhancing the expression of the *psbA* gene after high light stress, i.e. under increased amounts of oxidised glutathione. Enhancement of *psbA* gene expression should help to replenish D1 protein under high light intensities'

Finally, a recent in vivo study with transgenic *Arabidopsis* plants harbouring an *Apx*::*Luc* fusion suggests that the expression of the transgene can be influenced by reactive oxygen species (ROS), such as H_2O_2, in darkness. Thus, H_2O_2 can replace high light intensities and is claimed to act as a second messenger that allows intercellular communication, producing systemic acclimation. In this context, it is interesting to note that the H_2O_2-generating chloroplast superoxide dismutase from mustard can be purified as a complex with the plastid RNA polymerase A.

The reason for such an association is not yet clear, but there is definitely a need for efficient protection of newly synthesized transcripts against ROS generated by the photosynthetic electron transport. Therefore, it is possible that the H_2O_2 produced by the action of this enzyme could act as a signal that has an impact on chloroplast transcription.

CHLOROPLASTS, THE PHOTOSYNTHETIC UNITS OF GREEN PLANTS

The process of plant photosynthesis takes place entirely within the chloroplasts. Detailed studies of the role of these organelles date from the work of the British biochemist Robert Hill. About 1940 Hill discovered that green particles obtained from broken cells could produce oxygen from water in the presence of light and a chemical compound, such as ferric oxalate, able to serve as an electron acceptor.

This process is known as the Hill reaction. During the 1950s Daniel Arnon and other American biochemists prepared plant cell fragments in which not only the Hill reaction but also the synthesis of the energy-storage compound ATP occurred. In addition, the coenzyme NADP was used as the final acceptor of electrons, replacing the nonphysiological electron acceptors used by Hill.

His procedures were refined further so that individual small pieces of isolated chloroplast membranes, or lamellae, could perform the Hill reaction. These small pieces of lamellae were then fragmented into pieces so small that they performed only the light reactions of the photosynthetic process. It is now possible also to isolate the entire chloroplast so that it can carry out the complete process of photosynthesis, from light absorption, oxygen formation, and the reduction of carbon dioxide to the formation of glucose and other products.

Structural Features

The intricate structural organization of the photosynthetic apparatus is essential for the efficient performance of the complex process of photosynthesis. The chloroplast is enclosed in a double outer membrane, and its size approximates a spheroid about 2,500 nanometres thick and 5,000 nanometres long. Some single-celled algae have one chloroplast that occupies more than half the cell volume. Leaf cells of higher plants contain many chloroplasts, each approximately the size of the one in some algal cells.

When thin sections of a chloroplast are examined under the electron microscope, several features are apparent. Chief among these are the intricate internal membranes (*i.e.,* the lamellae) and the stroma, a colourless matrix in which the lamellae are embedded. Also visible are starch granules, which appear as dense bodies.

The stroma is basically a solution of enzymes and small molecules. The dark reactions occur in the stroma, the soluble enzymes of which catalyze the conversion of carbon dioxide and minerals to carbohydrates and other organic compounds. The capacity for carbon fixation and reduction is lost if the outer membrane of the chloroplast is broken, allowing the stroma enzymes to leak out.

A single lamella, which contains all the photosynthetic pigments, is approximately 10–15 nanometres thick. The lamellae exist in more-or-less flat sheets, a few of which extend through much of the length of the chloroplast. Examination of cross sections of lamellae under the electron microscope shows that their edges are joined to form closed hollow disks that are called thylakoids ("saclike"). The chloroplasts of most higher plants have regions, called grana, in which the thylakoids are very tightly stacked. When viewed by electron microscopy at an oblique angle, the grana appear as stacks of disks. When viewed in cross section, it is apparent that some thylakoids extend from one grana through the stroma into other grana. The thin aqueous spaces inside the thylakoids are believed to be connected with each other via these stroma thylakoids. These thylakoid spaces are isolated from the stroma spaces by the relatively impermeable lamellae.

The light reactions occur exclusively in the thylakoids. The complex structural organization of lamellae is required for proper thylakoid function; intact thylakoids apparently are necessary for the formation of ATP. Thylakoids that have been broken down to smaller units can no longer form ATP, even when the conversion of light into chemical energy occurs during electron transport in these units. Such lamellar fragments can carry out the Hill reaction, with the transfer of electrons from water to NADP.

Chemical composition of lamellae » Lipids

Lamellae consist of about equal amounts of lipids and proteins. About one-fourth of the lipid portion of the lamellae consists of pigments and coenzymes; the remainder consists of various lipids, including polar compounds such as phospholipids and galactolipids. These polar lipid molecules have "head" groups that attract water (*i.e.,* are hydrophilic) and fatty acid "tails" that are oil soluble and repel water (*i.e.,* are hydrophobic). When polar lipids are placed in an aqueous environment, they can line up with the fatty acid tails side by side. A second layer of phospholipids forms tail-to-tail with the first, establishing a lipid bilayer in which the hydrophilic heads are in contact with the aqueous solution on each side of the bilayer. Sandwiched between the heads are the hydrophobic tails, creating a hydrophobic environment from which water is excluded. This lipid bilayer is an essential feature of all biological membranes. The hydrophobic parts of proteins and lipid-soluble cofactors and pigments are dissolved or embedded in the lipid bilayer. Lamellar membranes can function as electrical insulating material and permit a charge, or potential difference, to develop across the membrane. Such a charge can be a source of chemical or electrical energy.

Approximately one-fifth of the lamellar lipids are chlorophyll molecules; one type, chlorophyll *a,* is more abundant than the second type, chlorophyll *b*. The chlorophyll molecules are specifically bound to small protein molecules. Most of these chlorophyll-proteins are "light-harvesting" pigments. These absorb light and pass its energy on to special chlorophyll *a* molecules that are directly involved in the conversion of light energy to chemical energy. When one of these special chlorophyll *a* molecules is excited by light energy (as described later), it gives up an electron. There are two types of these special chlorophyll *a* molecules: one, called P_{680}, has an absorption spectrum that peaks at 684 nanometres; the other, called P_{700}, shows an absorption peak at 700 nanometres.

Although chlorophylls are the main light-absorbing molecules in green plants, other pigments such as carotenes and carotenoids (which are responsible for the yellow-orange colour of carrots) also can absorb light and may supplement chlorophyll as the light-absorbing molecules in some plant cells. The light energy absorbed by these pigments must be passed to chlorophyll before conversion to chemical energy can occur.

Chemical Composition of Lamellae: Proteins

Many of the lamellar proteins are components of the chlorophyll–protein complexes described above. Other proteins include enzymes and protein-containing coenzymes. Enzymes are required as organic catalysts for specific reactions within the lamellae. Protein coenzymes, also called cofactors, include important electron carrier molecules called cytochromes, which are iron-containing pigments with the pigment portions attached to protein molecules. During electron transfer, an electron is accepted by an iron atom in the pigment portion of a cytochrome molecule, which thus is reduced; then the electron is transferred to the iron atom in the next cytochrome carrier in the electron transfer chain, thus oxidizing the first cytochrome and reducing the next one in the chain.

In addition to the metal atoms found in the pigment portions of cytochrome molecules, metal atoms also are found in other protein molecules of the lamellae. In proteins with a total molecular weight of 900,000 (based on the weight of hydrogen as one), there are two atoms of manganese, 10 atoms of iron, and six atoms of copper. These metal atoms are required for the catalytic activity of some of the enzymes important in photosynthesis. The manganese atoms are involved in water-splitting and oxygen formation. Both copper- and iron-containing proteins function in electron transport between water and the final electron-acceptor molecule of the light stage of photosynthesis, an iron-containing protein called ferredoxin. Ferredoxin is a soluble component in the chloroplasts. In its reduced form, it gives electrons directly to the systems that reduce nitrate and sulfate and via NADPH to the system that reduces carbon dioxide. A copper-containing protein called plastocyanin (PC) carries electrons at one point in the electron transport chain. PC molecules are water soluble and can move through the inner space of the thylakoids, carrying electrons from one place to another.

Chemical Composition of Lamellae: Quinones

Small molecules called plastoquinones are found in substantial numbers in the lamellae. Like the cytochromes, quinones have important roles in carrying electrons between the components of the light reactions. Since they are lipid soluble, they can diffuse through the membrane. They can carry one or two electrons and, in their reduced form (with added electrons), they carry hydrogen atoms that can be released as hydrogen ions when the added electrons are passed on, for example, to a cytochrome.

Light absorption and energy transfer

The light energy absorbed by a chlorophyll molecule excites some electrons within the structure of the molecule to higher energy levels, or excited states. Light of shorter wavelength has more energy than light of longer wavelength , so that absorption of blue light creates an excited state of higher energy. A molecule raised to this higher energy state quickly gives up the "extra" energy as heat and falls to its lowest excited state. This lowest excited state is similar to that of a molecule that has just absorbed the longest wavelength light capable of exciting it. In the case of chlorophyll *a*, this lowest excited state corresponds to that of a molecule that has absorbed red light of about 680 nanometres. the return of a chlorophyll *a* molecule from its lowest excited state to its original low-energy state (ground state) requires the release of the extra energy of the excited state.

This can occur in one of several ways. In photosynthesis, most of this energy is conserved as chemical energy by the transfer of an electron from a special chlorophyll *a* molecule (P_{680} or P_{700}) to an electron acceptor. When this electron transfer is blocked by inhibitors, such as the herbicide dichlorophenylmethylurea (DCMU), or by low temperature, the energy can be released as red light. Such re-emission of light is called fluorescence. The examination of fluorescence from photosynthetic material in which electron transfer has been blocked has proved to be a useful tool for scientists studying the light reactions.

SYNTHESIS OF ATP BY THE ATP SYNTHASE ENZYME

The conversion of proton electrochemical energy into chemical free energy is accomplished by a single protein complex known as ATP synthase. This enzyme catalyzes a phosphorylation reaction, which is the formation of ATP by the addition of inorganic phosphate (Pi) to ADP

$$ADP–3 + Pi\text{-}2 + H+ \rightarrow ATP–4 + H_2O. \quad (5)$$

The reaction is energetically uphill (DG = +32 kJ/mol) and is driven by proton transfer through the ATP synthase protein. The ATP Synthase complex is composed of two major subunits, CF0 and CF1.

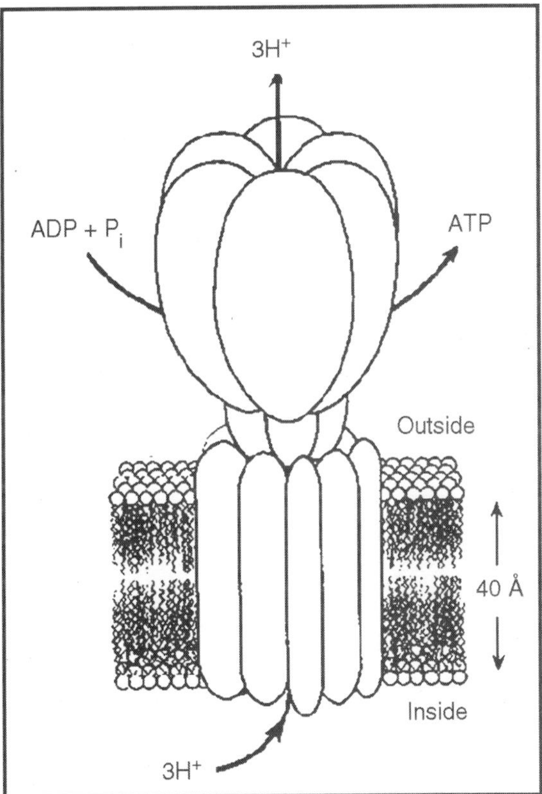

The CF0 subunit spans the photosynthetic membrane and forms a proton channel through the membrane. The CF1 subunit is attached to the top of the CF0 on the outside of the membrane and is located in the aqueous space. CF1 is composed of several different protein subunits, referred to as a, b, g, d and e.

The top portion of the CF1 subunit is composed of three ab-dimers that contain the catalytic sites for ATP synthesis. A recent major breakthrough has been the elucidation of the structure of ATPase of beef heart mitochondria by Abrahams et al. (1994).

The molecular processes that couple proton transfer through the protein to the chemical addition of phosphate to ADP are poorly understood. It is known that phosphorylation can be driven by a pH gradient, a transmembrane electric field, or a combination of the two. Experiments indicate that three protons must pass through the ATP synthase complex for the synthesis of one molecule of ATP.

However, the protons are not involved in the chemistry of adding phosphate to ADP. Paul Boyer and coworkers have proposed an alternating binding site mechanism for ATP synthesis. One model based on their proposal is that there are three catalytic sites on each CF1 that cycle among three different states.

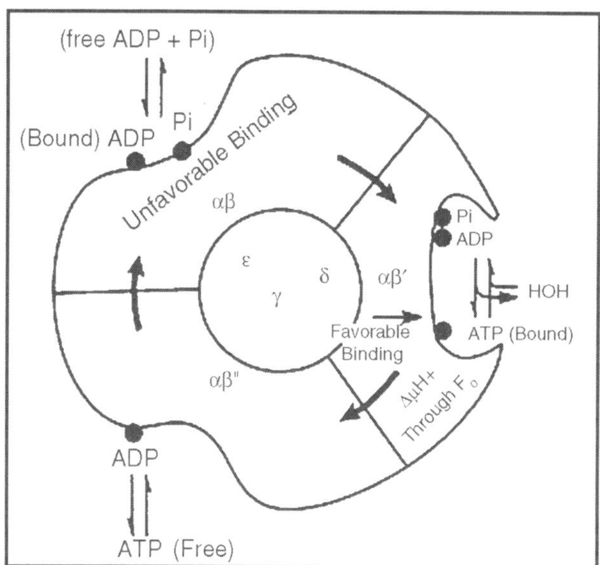

The states differ in their affinity for ADP, Pi and ATP. At any one time, each site is in a different state. This model is supported by the structure of ATPase elucidated by Abrahams et al.. Initially, one catalytic site on CF1 binds one ADP and one inorganic phosphate molecule relatively loosely.

Due to a conformational change of the protein, the site becomes a tight binding site, that stabilizes ATP. Next, proton transfer induces an alteration in protein conformation that causes the site to release the ATP molecule into the aqueous phase. In this model, the energy from the proton electrochemical gradient is used to lower the affinity of the site for ATP, allowing its release to the water phase.

The three sites on CF1 act cooperatively, i.e., the conformational states of the sites are linked. It has been proposed that protons affect the conformational change by driving the rotation of the top part (the three ab-dimers) of CF1.

Such a rotating model has recently been supported by recording of a rotation of the gamma subunit relative to the alpha-beta subunits by Sabbert et al.. This revolving site mechanism would require rates as high as 100 revolutions per second. It is worth noting that flagella that propel some bacteria are driven by a proton pump and can rotate at 60 revolutions per second.

Synthesis of Carbohydrates

All plants and algae remove CO2 from the environment and reduce it to carbohydrate by the Calvin cycle. The process is a sequence of biochemical reactions that reduce carbon and rearrange bonds to produce carbohydrate from CO2 molecules.

The first step is the addition of CO2 to a five-carbon compound (ribulose 1,5-bisphosphate). The six-carbon compound is split, giving two molecules of a three-carbon compound (3-phosphoglycerate). This key reaction is catalyzed by Rubisco, a large water soluble protein complex.

The 3-dimensional structure has been determined by X-ray analysis for Rubisco isolated from tobacco from a cyanobacterium (Synechococcus) and from a purple bacterium (Rhodospirillum rubrum. The carboxylation reaction is energetically downhill.

The main energy input in the Calvin cycle is the phosphorylation by ATP and subsequent reduction by NADPH of the initial three-carbon compound forming a three-carbon sugar, triosephosphate. Some of the triosephosphate is exported from the chloroplast and provides the building block for synthesizing more complex molecules. In a process known as regeneration, the Calvin cycle uses some of the triosephosphate molecules to synthesize the energy rich ribulose 1,5-bisphosphate needed for the initial carboxylation reaction.

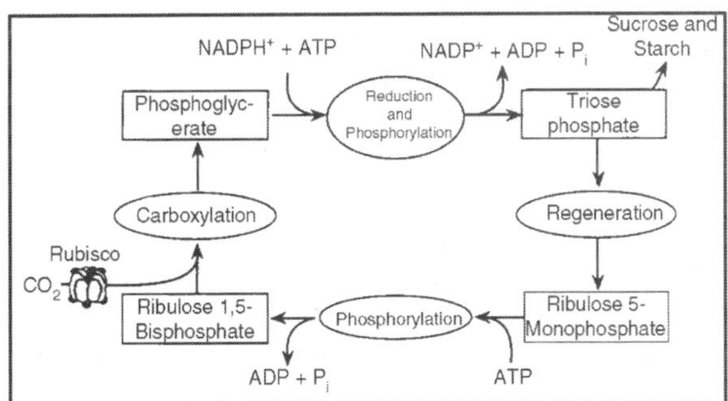

This reaction requires the input of energy in the form of one ATP. Overall, thirteen enzymes are required to catalyze the reactions in the Calvin cycle. The energy conversion efficiency of the Calvin cycle is approximately 90%. The reactions do not involve energy transduction, but rather the rearrangement of chemical energy. Each molecule of CO_2 reduced to a sugar $[CH_2O]n$ requires 2 molecules of NADPH and 3 molecules of ATP.

Rubisco is a bifunctional enzyme that, in addition to binding CO_2 to ribulose bisphosphate, can also bind O_2. This oxygenation reaction produces the 3-phosphoglycerate that is used in the Calvin cycle and a two-carbon compound (2-phosphoglycolate) that is not useful for the plant.

In response, a complicated set of reactions (known as photorespiration) are initiated that serve to recover reduced carbon and to remove phosphoglycolate. The Rubisco oxygenation reaction appears to serve no useful purpose for the plant. Some plants have evolved specialized structures and biochemical pathways that concentrate CO_2 near Rubisco. These pathways (C4 and CAM), serve to decrease the fraction of oxygenation reactions.

PHOTOSYSTEM II AND PHOTOSYSTEM I REACTION CENTERS

Photosystem II uses light energy to drive two chemical reactions - the oxidation of water and the reduction of plastoquinone. The photosystem II complex is composed of more than fifteen polypeptides and at least nine different redox components (chlorophyll, pheophytin, plastoquinone, tyrosine, Mn, Fe, cytochrome b559, carotenoid and histidine) have been shown to undergo light-induced electron transfer. However, only five of these redox components are known to be involved in transferring electrons from H_2O to the plastoquinone pool - the water oxidizing manganese cluster $(Mn)_4$, the amino acid tyrosine, the reaction centre chlorophyll (P680), pheophytin, and the plastoquinone molecules, QA and QB. Of these essential redox components, tyrosine, P680, pheophytin, QA and QB have been shown to be bound to two key polypeptides that form the heterodimeric reaction centre core of photosystem II (D1 and D2).

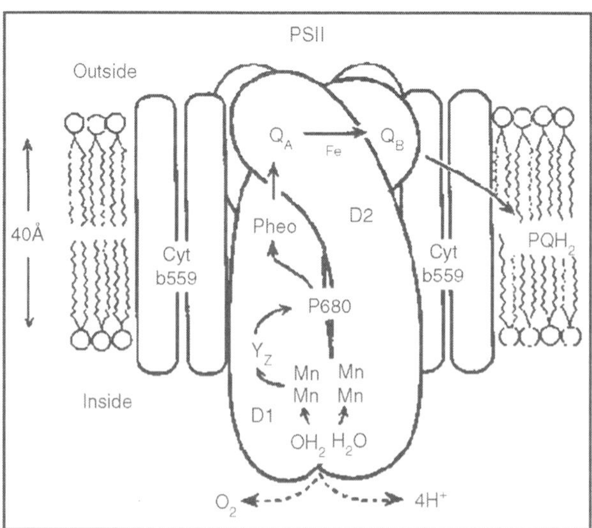

Recent work indicates that the D1 and D2 polypeptides also provide ligands for the (Mn)4 cluster. The three-dimensional structure of photosystem

II is not known. Our knowledge of its structure is guided by the known structure of the reaction centre in purple bacteria and biochemical and spectroscopic data. Fig. shows a schematic view of photosystem II that is consistent with current data.

Photochemistry in photosystem II is initiated by charge separation between P680 and pheophytin, creating P680+/Pheo-. Primary charge separation takes about a few picoseconds.

Subsequent electron transfer steps have been designed through evolution to prevent the primary charge separation from recombining. This is accomplished by transferring the electron within 200 picoseconds from pheophytin to a plastoquinone molecule (QA) that is permanently bound to photosystem II. Although plastoquinone normally acts as a two-electron acceptor, it works as a one-electron acceptor at the QA-site. The electron on QA- is then transferred to another plastoquinone molecule that is loosely bound at the QB-site. Plastoquinone at the QB-site differs from QA in that it works as a two-electron acceptor, becoming fully reduced and protonated after two photochemical turnovers of the reaction centre.

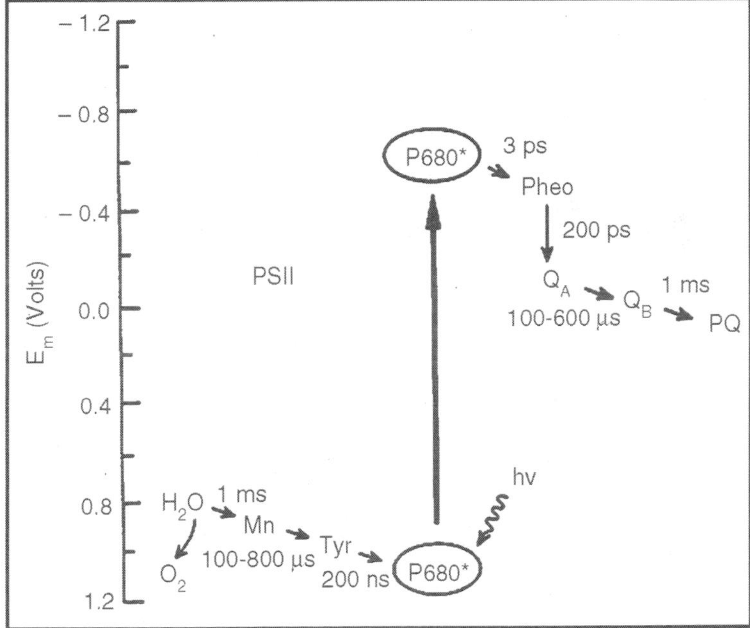

The full reduction of plastoquinone requires the addition of two electrons and two protons, i.e., the addition of two hydrogen atoms. The reduced plastoquinone then debinds from the reaction centre and diffuses into the hydrophobic core of the membrane. After which, an oxidized plastoquinone molecule finds its way to the QB-binding site and the process is repeated. Because the QB-site is near the outer aqueous phase, the protons added to plastoquinone during its reduction are taken from the outside of the membrane.

Plastoquinone
(reduced form)

H
|
O
H_3C H
CH_3
H_3C $(CH_2-CH=\overset{|}{C}-CH_2)_9H$
O
|
H

Photosystem II is the only known protein complex that can oxidize water, resulting in the release of O2 into the atmosphere. Despite years of research, little is known about the molecular events that lead to water oxidation. Energetically, water is a poor electron donor. The oxidation- reduction midpoint potential (Em,7) of water is +0.82 V (pH 7). In photosystem II this reaction is driven by the oxidized reaction centre, P680+ (the midpoint potential of P680/P680+ is estimated to be +1.2 V at pH 7).

How electrons are transferred from water to P680+ remains a mystery. It is known that P680+ oxidizes a tyrosine on the D1 protein and that Mn plays a key role in water oxidation. Four Mn ions are present in the water oxidizing complex. X-ray absorption spectroscopy shows that Mn undergoes light-induced oxidation.

Water oxidation requires two molecules of water and involves four sequential turnovers of the reaction centre. This was shown by an experiment demonstrating that oxygen release by photosystem II occurs with a four flash dependence. Each photochemical reaction creates an oxidant that removes one electron. The net reaction results in the release of one O2 molecule, the deposition of four protons into the inner water phase, and the transfer of four electrons to the QB-site (producing two reduced plastoquinone molecules)

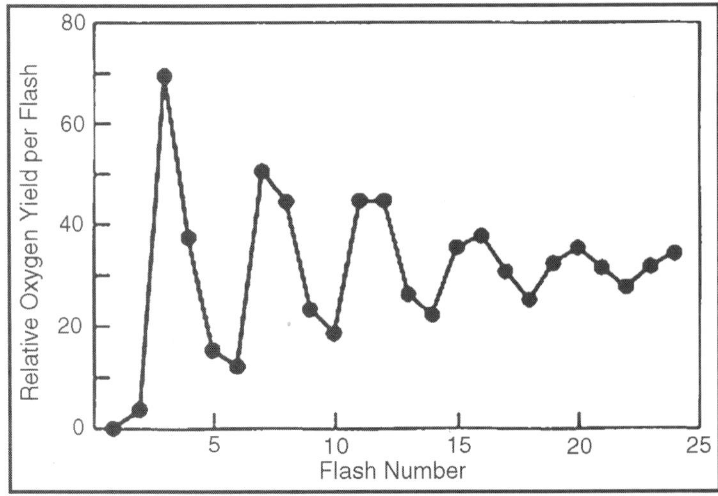

Photosystem II reaction centers contain a number of redox components with no known function. An example is cytochrome b559, a heme protein, that is an essential component of all photosystem II reaction centers. If the cytochrome is not present in the membrane, a stable PS II reaction centre cannot be formed.

Although the structure and function of Cyt b559 remain to be discovered, it is known that the cytochrome is not involved in the primary enzymatic activity of PS II, which is the transfer of electrons from water to plastoquinone. Why PS II reaction centers contain redox components that are not involved in the primary enzymatic reactions is a puzzling question. The answer may be found in the unusual chemical reactions occurring in PS II and the fact that the reaction centre operates at a very high power level.

Photosystem II is an energy transforming enzyme that must switch between various high energy states that involve the creation of the powerful oxidants required for removing electrons from water and the complex chemistry of plastoquinone reduction which is strongly influenced by protons. In saturating light a single reaction centre can have an energy throughput of 600 eV/s. Operating at such a high power level results in damage to the reaction centre. It may be that some of the "extra" redox components in photosystem II may serve to protect the reaction centre.

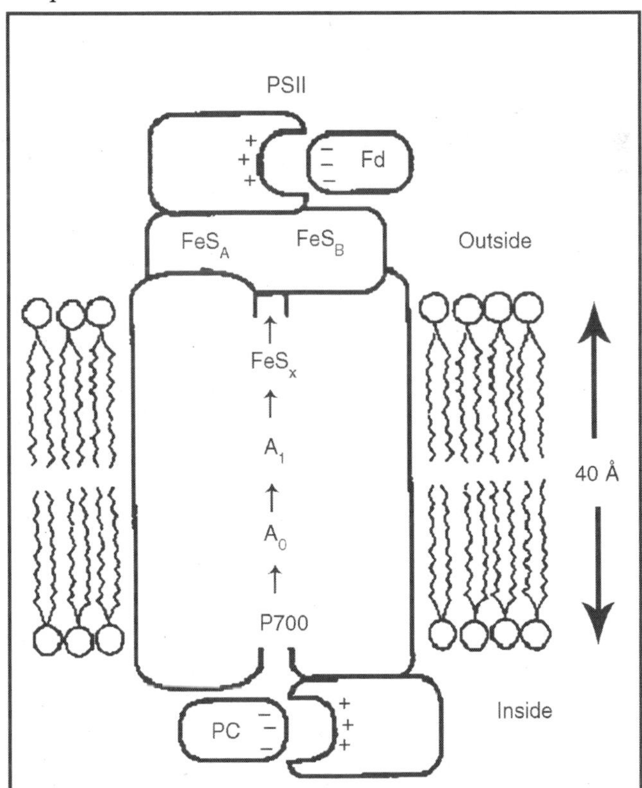

Photosystem II has another perplexing feature. Many plants and algae have been shown to have a significant number of photosystem II reaction centers that do not contribute to photosynthetic electron transport. Why plants devote resources for the synthesis of reaction centers that apparently do not contribute to energy conversion is unknown. The photosystem I complex catalyzes the oxidation of plastocyanin, a small soluble Cu- protein, and the reduction of ferredoxin, a small FeS protein. Photosystem I is composed of a heterodimer of proteins that act as ligands for most of the electron carriers.

The reaction centre is served by an antenna system that consists of about two hundred chlorophyll molecules (mainly chlorophyll a) and primary photochemistry is initiated by a chlorophyll a dimer, P700. In contrast to photosystem II, many of the antenna chlorophyll molecules in photosystem I are bound to the reaction centre proteins.

Also, FeS centers serve as electron carriers in photosystem I and, so far as is known, photosystem I electron transfer is not coupled to proton translocation. Primary charge separation occurs between a primary donor, P700, a chlorophyll dimer, and a chlorophyll monomer (Ao). The subsequent electron transfer events and rates are shown in Figure.

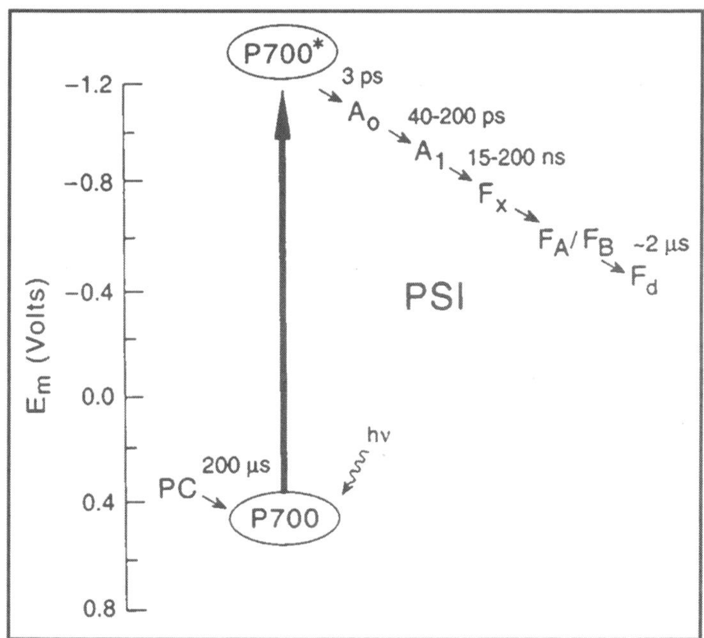

Electron Transport

Electron transport from water to NADP requires three membrane bound protein complexes operating in series - photosystem II, the cytochrome bf complex and photosystem I. Electrons are transferred between these large protein complexes by small mobile molecules (plastoquinone and plastocyanin in plants). Because these small molecules carry electrons (or hydrogen atoms) over relatively long distances, they play a unique role in photosynthetic energy conversion. This is illustrated by plastoquinone (PQ), which serves two key functions.

Plastoquinone transfers electrons from the photosystem II reaction centre to the cytochrome bf complex and carries protons across the photosynthetic membrane. It does this by shuttling hydrogen atoms across the membrane from photosystem II to the cytochrome bf complex. Because plastoquinone is hydrophobic its movement is restricted to the hydrophobic core of the photosynthetic membrane.

Plastoquinone operates by diffusing through the membrane until, due to random collisions, it becomes bound to a specific site on the photosystem II complex. The photosystem II reaction centre reduces plastoquinone at the QB-site by adding two electrons and two protons creating PQH2. The reduced plastoquinone molecule debinds from photosystem II and diffuses randomly in the photosynthetic membrane until it encounters a specific binding site on the cytochrome bf complex. The cytochrome bf complex is a membrane bound protein complex that contains four electron carriers, three cytochromes and an FeS centre.

The crystal structure has been solved for cytochrome f from turnip and the FeS centre from bovine heart mitochondria. In a complicated reaction sequence that is not fully understood, the cytochrome bf complex removes the electrons from reduced plastoquinone and facilitates the release of the protons into the inner aqueous space. The electrons are eventually transferred to the photosystem I reaction centre. The protons released into the inner aqueous space contribute to the proton chemical free energy across the membrane.

Electron transfer from the cytochrome bf complex to photosystem I is mediated by a small Cu-protein, plastocyanin (PC). Plastocyanin is water soluble and operates in the inner water space of the photosynthetic membrane. Electron transfer from photosystem I to NADP requires ferredoxin, a small FeS protein, and ferredoxin-NADP oxidoreductase, a peripheral flavoprotein that operates on the outer surface of the photosynthetic membrane. Ferredoxin and NADP are water soluble and are found in the outer aqueous phase.

The pathway of electrons is largely determined by the energetics of the reaction and the distance between the carriers. The electron affinity of the carriers is represented in Figure. by their midpoint potentials, which show the free energy available for electron transfer reactions under equilibrium conditions. (It should be kept in mind that reaction conditions during photosynthesis are not in equilibrium.)

Subsequent to primary charge separation, electron transport is energetically downhill (from a lower (more negative) to a higher (more positive) redox potential). It is the downhill flow of electrons that provides free energy for the creation of a proton chemical gradient.

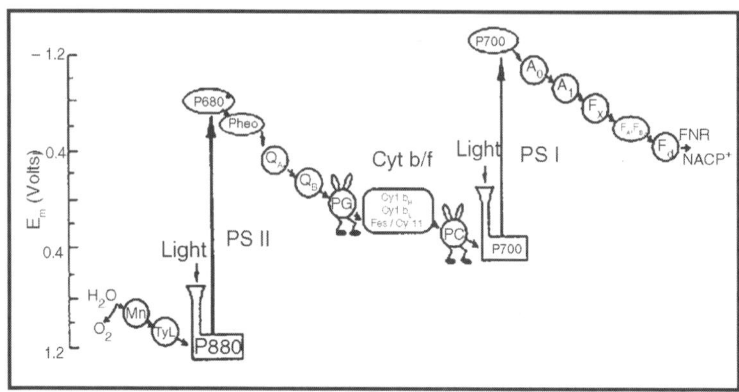

Photosynthetic membranes effectively limit electron transport to two dimensions. For mobile electron carriers, limiting diffusion to two dimensions increases the number of random encounters. Furthermore, because plastocyanin is mobile, any one cytochrome bf complex can interact with a number of photosystem I complexes. The same is true for plastoquinone, which commonly operates at a stoichiometry of about six molecules per photosystem II complex.

ORGANIZATION OF ENERGY TRANSFER NETWORKS IN PHOTOSYNTHESIS

Photosynthesis is the primary energy source for most of the biosphere. Even though photosynthetic light harvesting complexes display great variety in their design and function, many of them are membrane proteins that comprise of pigment antenna arrays which absorb light and transfer the resulting electronic excitation to a reaction centre which in turn converts this excitation energy to a charge gradient across the membrane.

DRIVE TOWARDS COMPLEXITY

It is of interest to compare anoxygenic light harvesting mechanisms to oxygenic ones. Purple bacteria, which perform anoxygenic photosynthesis, feature the oldest known light harvesting apparatus, while the oxygenic cyanobacteria, plants and green algea are evolutionarily more recent. In a purple bacterial photosynthetic unit pigments form highly symmetric ring-like structures. In contrast, the pigment network of photosystem I from oxygenic species forms a rather random looking array with a higher packing density of chlorophyll per unit mass.

A comparison of anoxygenic and oxygenic light harvesting mechanisms. (left) A sample photosynthetic unit (PSU) of purple bacteria: a peripheral light harvesting complex LH2 (small ring) funnels its harvested light energy to an LH1 complex (big ring) which in turn delivers its energy to the reaction centre it surrounds. There might be up to around ten LH2's per LH1 in a PSU. (right) Photosystem I from cyanobacteria combine the peripheral antenna array, the reaction centre, and the electron transfer chain in to a single protein. The circular arrangement of pigments found in purple bacteria seems to be a consequence of simple assembly and aggregation requirements rather than having a specific functional role, since the effects of thermal disorder erase any symmetry artifacts that would be left over if the system were to function at cryogenic temperatures. This raises the question of whether the more closely packed and seemingly random chlorophyll network, found in cyanobacteria and plants, serves a particular functional role or whether it can be safely regarded as a 'random bag of chlorophylls'. Are there overarching organizational principles for the geometry of chlorophyll networks? In order to address this question the kinetics of the excitation migration process, which takes place in a picosecond time scale, needs to studied in some detail.

From Structure to Function

The physics of the excitation migration process is fairly well understood and is described to a reasonable approximation by Foerster theory within the context of an effective Hamiltonian formalism. This framework enables one to construct, from structural information regarding cholorophyll position and orientations such as provided by the work of our collaborator Petra Fromme, the geometry of an excitation transfer network as well as stochastic formulations for the measurable quantities of the excitation transfer process such as the average excitation lifetime or the overall quantum yield. (Quantum yield is the probability with which the energy of an absorbed photon is utilized for charge transfer as opposed to being lost to thermal dissipation.) Remarkable agreement between the computed timescales and quantum yields with observed ones indicate that the fundamentals of the excitation migration are understood at least to some reasonable degree.

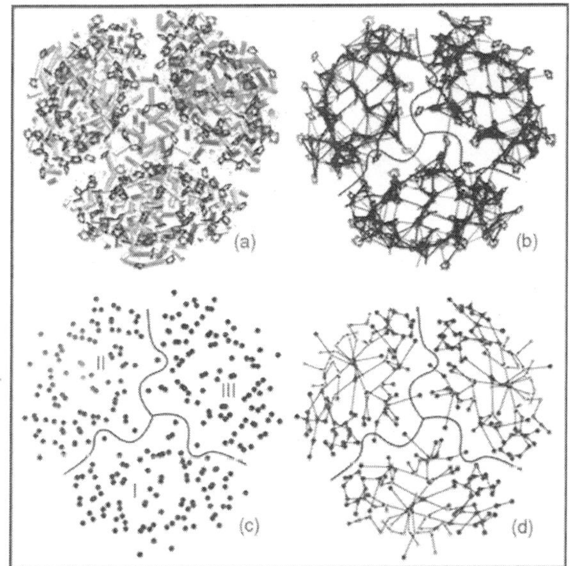

Four ways of looking at the chlorophyll network of the photosystem I trimer. (a) The protein—pigment complex (cofactors other than chlorophylls not shown). (b) Excitation transfer rates between individual chlorophylls. The thickness of a bond between two chlorophylls is proportional to the logarithm of the transfer rate between them. For simplicity, only the largest rates are shown. (c) Connectivity between different monomers. The measure is given by the probabilities of charge transfer from a given reaction centre for an initial condition corresponding to a perfectly localized excitation at a given chlorophyll. The content of red, green, and blue are proportional to the probability of charge transfer from the bottom, left and right reaction centers, respectively. (d) Representative pathways of excitation migration. The arrows between chlorophylls characterizing the migration are assigned according to a steepest-descent criterion based on mean first passage times to the reaction centre. The colors denote increasing mean first passage times from red to blue.

However, an analysis of the light harvesting function based on the frozen geometry of an atomic structure is not a sufficient description for a system that functions at physiological temperatures.

Thermal Disorder

The processes of photon absorption, excitation migration and electron transfer are all fundamentally quantum mechanical in nature. Since these processes occur at finite temperature, a satisfactory description of the light harvesting function needs to account for the effects of thermal disorder in a quantum mechanical system.

Random matrix theories, widely employed in condensed matter physics and high energy physics, provide a framework within which thermal effects can be formulated in terms of ensemble averages over a space of random Hamiltonians. This enables one to employ field theory methodologies for finding elegant and succint descriptions, for example, for spectral properties of a light harvesting complex at nonzero temperatures.The theoretical framework outlined so far can now be used to compare various chlorophyll network geometries such as ones that arise from thermal fluctuations, external perturbations (such as loss of a chlorophyll), or a competitive new design of a neighboring species. How much change can a chlorophyll network take and still survive?

Surviving Thermal Disorder, Environmental Change and Competition

Survivability. What is a good quantitative measure for survivability? What are the most important evolutionary pressures that a species has to contend with?

Environmental change and competition are two major challenges that all biological systems must cope with. Adaptability to changing external conditions, or robustness, of a system typically manifests itself in terms of a parameter insensitivity of its dynamics and a graceful degradation of its components. Competition, on the other hand, drives a system towards optimality, as a less efficient system will find itself at an evolutionary disadvantage.

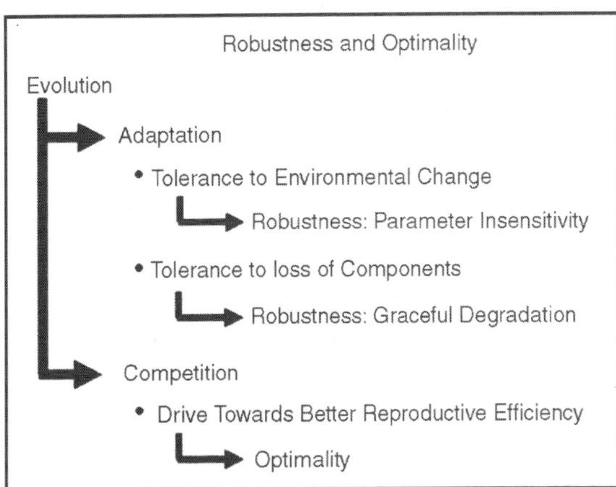

Adaptive and competitive pressures of evolution drive a species toward tolerance to change, i.e. robustness, on the one hand and toward high relative efficiency, i.e. optimality, on the other.

It is very difficult to quantify robustness and optimality in general terms for an arbitrary biological system since the fitness landscape over which adaptability needs to be judged is enormously complex. A light harvesting system, however, provides a natural, if somewhat crude, measure of its efficiency in terms of the quantum yield of the excitation migration process.

It is a simple matter to model the effects of various perturbations, such as thermal disorder or loss of individual components, on the quantum yield within the theoretical framework outlined above. Similarly, questions regarding the optimality of the geometry of the chlorophyll network can be investigated by generating ensembles of alternative network configurations.

Quantum Yield of a Light Harvesting System as a (crude) Measure of Fitness

The quantum yield of a light harvesting system is typically very high. It is given by a near unit probability, as most every photon absorbed by the chlorophyll network results in an electron transfer. Strangely, this no longer remains true at cryogenic temperatures, where quantum yield drops significantly and becomes dependent on the wavelength of the incident photon. This results from the loss of spectral overlap necessary for resonant energy transfer between donor and acceptor chlorophylls that may have different energies.

(Having all pigments to have the same energy would have solved the resonance problem at any temperature, but then only a tiny window of the spectrum would be covered by the pigment array, save for spectral broadening due to excitonic coupling.) Thus, thermal disorder actually helps to maintain a high quantum yield by broadening cholorophyll spectral lineshapes and enabling resonant energy transfer even with some energy separation between chlorophylls.

A study of quantum yields in an ensemble of randomly generated chlorophyll network geometries show that the high quantum yield of the system remains manifest under thermal fluctuations of chlorophyll orientations, chlorophyll site energies, as well as even loss of individual chlorophylls from the network. Thus, there is a degree of robustness present in such a chlorophyll network as quantified by the quantum yield.

However, within this narrow range of quantum yields for random geometries it is seen that the original configuration, as given by the geometry corresponding to the crystal structure, is nearly at the top of the distribution.

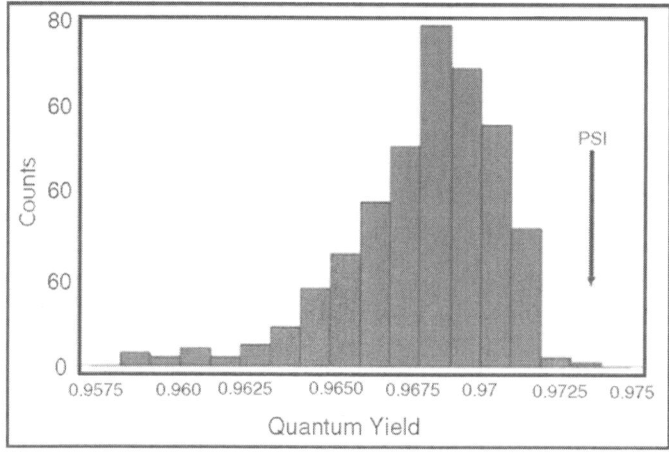

High relative quantum yield as a sign of optimality. Histograms of quantum yield are shown within an ensemble of random chlorophyll reorientations. The quantum yield of the original geometry given by the known atomic structure is indicated by an arrow. (top) *All* chlorphylls randomly reoriented, including reaction centre chlorophylls. (bottom) Six reaction centre chlorophylls kept fixed while the rest are randomly reoriented.

The high relative quantum yield of the original geometry within an ensemble of random configurations can be regarded as a sign of optimality. However, this raises the question of whether a quantum yield increase of a fraction of per cent is of any importance for the survival of a species. It is unlikely that for the day-to-day functioning of a light harvesting complex such a tiny change will be noticable, but over evolutionary timescales the cumulative effect of a slightly better efficiency can be a discriminating factor, as suggested also by growth competition experiments.

It must also be kept in mind that excitation transfer is not a *rate limiting* step for the overall light harvesting process, therefore its efficiency as given by the quantum yield can provide only a crude and incomplete measure for fitness. It still remains a challenge today to formulate other quantitative measures for the robustness and optimality of a light harvesting system that are still computationally tractable.

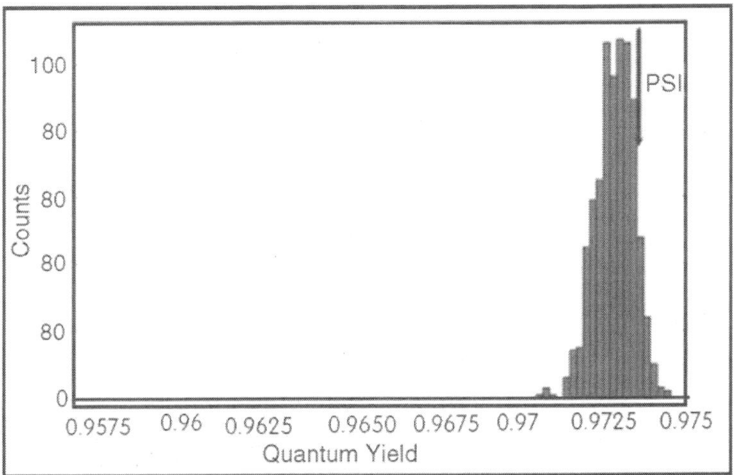

Index